W9-CXU-942

Green Electronics Design and Manufacturing

Implementing Lead-Free and RoHS-Compliant Global Products

Sammy G. Shina

New York Chicago San Francisco
Lisbon London Madrid Mexico City
Milan New Delhi San Juan
Seoul Singapore Sydney Toronto

The McGraw·Hill Companies

Cataloging-in-Publication Data is on file with the Library of Congress.

McGraw-Hill books are available at special quantity discounts to use as premiums and sales promotions, or for use in corporate training programs. To contact a representative please visit the Contact Us pages at www. mhprofessional.com.

Green Electronics Design and Manufacturing

1 2 3 4 5 6 7 8 9 0   DOC/DOC   0 1 4 3 2 1 0 9 8

ISBN 978-0-07-149594-3
MHID 0-07-149594-0

This book was printed on recycled, acid-free paper containing a minimum of 50% recycled, de-inked fiber.

Sponsoring Editor	**Proofreader**
Stephen S. Chapman	Malvika Shyam
Acquisitions Coordinator	**Indexer**
Alexis Richard	WordCo Indexing Services
Editorial Supervisor	**Production Supervisor**
David E. Fogarty	Pamela Pelton
Project Manager	**Composition**
Vastavikta Sharma, International Typesetting and Composition	International Typesetting and Composition
Copy Editor	**Art Director, Cover**
Patti Scott	Jeff Weeks

To my wife, Jackie,
and our children and grandchildren.

About the Author

Sammy G. Shina, P.E., is a professor of mechanical engineering at the University of Massachusetts Lowell, and has previously lectured at the University of Pennsylvania's EXMSE Program and at the University of California, Irvine. He is a past chairman of the Society of Manufacturing Engineers (SME) Robotics/ FMS and a founding member of the Massachusetts Quality Award. He is the founder of the New England Lead-Free Consortium, with over 30 contributing companies which are engaged in electronics products and their supply chain since its inception in 1999. The consortium is actively researching, testing, and evaluating materials and processes for lead-free and RoHS compliance, and it is funded by TURI, EPA, and the member companies. The consortium has published over 40 papers; some were translated into Asian languages and won a regional EPA Environmental Merit Award for the business category in May 2006. Dr. Shina is the author of two best-selling books on concurrent engineering, and a Six Sigma book, and he authored two chapters and over 100 technical publications in his fields of research.

Dr. Shina is an international consultant, trainer, and seminar provider on lead-free, quality, Six Sigma, and DoE as well as project management, technology supply chains, product design and development, and electronics manufacturing and automation. He worked for 22 years in high-technology companies developing new products and state-of-the-art manufacturing technologies. He was the speaker for the HP Executive Seminars on Concurrent Product/Process Design, Mechanical CAD Design and Test, and the Motorola Six Sigma Institute. He received S.B. degrees in electrical engineering and industrial management from MIT, a S.M. degree in computer science from WPI, and a Sc.D. degree in mechanical engineering from Tufts University. He resides in Framingham, Massachusetts.

Contents

Illustrations

Tables

Preface

Green electronics design and manufacturing is becoming more important as companies compete in a worldwide market for customers who desire high-quality and low-cost products, yet want to be good environmentalists who would expend the least amount of energy and contribute the minimum amount of hazardous materials into the air we breathe, the water we drink or send into the ocean, and land we inhabit. Consumers have consistently indicated that they will pay a premium for products deemed more environmentally friendly or "green." Most surveys indicate that they would be willing to pay 10 percent extra if a product they purchased was green. Governments and politicians have responded to these popular movements and have started to legislate mandatory compliance with a variety of limitations and bans on the use of certain hazardous materials and chemicals.

The push for mandating green products has come initially from Europe and Japan. Both regions have dense populations and limited land area, especially in the case of Japan. In addition, home-grown green movements could be used as a competitive weapon or a source of trade barriers, as well as a nationalist or regional source of pride. In particular, the ROHS directive of July 2006 was a landmark decision by the European Union (EU) to mandate the banning of certain hazardous metals and chemicals. ROHS compliance was necessary for all products to be sold in the EU. The primary attention that was given to ROHS by the electronics industry was due to lead, which was one of the banned substances and is used in the attachment of electronic components through leaded solder. This directive quickly became a rallying effort by all countries and industries to convert to greener products, by including the elimination of other metals and chemicals from the design and manufacturing processes.

In the United States, the progress of green design and manufacturing has been much slower, mostly due to the fact that the United States considered itself to be the place where the modern electronics industry began, and the country was concerned that it was being pushed too hastily into the green movement. In the U.S.

industry viewpoint, it took a very long time to achieve high levels of quality and reliability in electronic products, and too quick a move into alternate materials and processes would jeopardize this successful and reliable stage of electronic products' quality.

Initial U.S. companies' reaction was to fight the hazardous materials ban, by pointing out either that these materials were not as bad, or that the replacement materials were just as bad. Another option to fight the green monster would be to recycle nongreen materials, or to better protect the workers handling them.

U.S. popular culture injected itself into the green discussions. Movies highlighting the health dangers of hazardous materials, such as "Silkwood" in 1983 about the handling of nuclear materials in Kerr McGee Corporation in Utah, "Civil Action" in 1998 about the Woburn, Massachusetts, case against WR Grace Company, and "Erin Brockovich" in 2000 about a California town fight against Pacific Gas and Electric, all presented the health dangers of hazardous materials in a raw, emotional, and powerful manner. In addition, new paradigms such as *global warming* and *carbon footprint* have taken hold on the culture to build a special and popular language code or "Lingua Franca" of green. Certainly, no U.S. company wanted to be seen as opposing something good and advocating injury to people or the environment.

The current state of green design and manufacturing is that companies are ready to implement green products and processes and in fact are anticipating greater reduction in the use of energy and hazardous materials. They are doing so by pooling their knowledge in consortia and associations and making sure to stay ahead of the curve against the next targeted materials. A good example of such an effort is the National Electronic Manufacturer Institute or iNEMI effort to eliminate the use of halogens in the printed-circuit board (PCB) laminates. This would be a good model for future directions in green design and manufacturing.

About the Book Organization

This book is intended to introduce and familiarize design, production, quality, and process engineers and their managers with many of the issues regarding the use of green materials and processes in the design and manufacturing of electronic products, and how to resolve them. It is based on the author's experience in researching, practicing, consulting, and teaching green materials and processes and their techniques in the past 10 years. During that time, I confronted many engineers' natural reservations about green materials: its bans are too arbitrary, it is too difficult to achieve, it works only for larger companies willing to invest a huge amount of effort and research into green, it is too expensive to implement, it is only for manufacturing, not for design, and so on. They continuously challenged me to apply it in their own areas of interest, presenting me with many difficult green design and manufacturing applications and problems to solve. At the same time, I was involved in many companies and organizations whose engineers and managers were using original and ingenious applications of green methodologies and materials in traditional design and manufacturing. Out of these experiences came many of the examples and case studies in this book.

One of the most important sources for the materials for this book is the work of the Massachusetts Lead-Free Consortium that I formed back in 1999. It started with two companies and grew quickly to eight companies. Then additional companies joined in from other New England states, and we changed the name to the New England Lead-Free Consortium. All the contributed chapters are by leading engineers and professionals from the consortium companies.

The companies participating in the consortium vary in size, scope, product type, and strategy. Yet they are similar in their approach by successfully implementing green design and manufacturing through an interdisciplinary team environment and using the tools and methods mentioned in this book effectively and by altering them to meet their particular needs.

I believe the most important impact of green design and manufacturing is to use the techniques and methods mentioned in this book in the early design phase of new products, starting with making green one of the goals of the new product creation process. It would make the design engineers extremely cognizant of the importance of designing and specifying products that can be manufactured with green materials and processes, and with lower cost and energy consumption. The ideal condition exists when a company introduces green products by having manufacturing and the supply chain fully commit and already experienced in supplying and manufacturing green materials and processes. The company and its suppliers should have a good sense of the cost of, quality of, and reliability associated with green. This is indeed a very daunting task, especially if the supply chain consists of smaller companies and is not yet fully prepared for green materials and process.

The approach I used in this book is not to be rigid about green design and manufacturing. I have attempted to give many of the options available to implement green materials and processes, and not to specifically recommend a course of action in each instance. Engineers are very creative people, and they will always try to meld new concepts into ones familiar to them. Many will put their own stamp on green products or add their own way of doing things to the design and manufacturing of such products. The one sure thing to make them resist a new concept is to force it down their throats. I believe these individual engineers' efforts should be encouraged, as long as they do not detract from the overall goal of achieving green products.

This book is divided up into two parts. Part I recounts the principles and tools necessary for initiating green products and processes. The tools include the use of statistical techniques for comparison of green alternative materials and processes versus nongreen baselines. Other important techniques for developing alternate materials and processes are the use of Design of Experiments (DoE) methodologies for selecting the most optimal green replacement for materials and processes. In addition, the principles of reliability and the use of different techniques to simulate life testing are presented. Actual examples of using these techniques for green design and manu-facturing are presented after discussions of the mathematical principles used. I wrote the chapters on statistics and DoE, and David Pinksy, a senior reliability engineer and expert on tin whiskers, wrote Chap. 3 on reliability.

Part II of the book is a comprehensive set of successful implementations of green products and processes. They range from the viewpoints of senior managers, to project managers, and to design and manufacturing experts. The green implementations were drawn from companies in the New England Lead-Free Consortium that have successfully managed a successful conversion to green products and processes.

I hope that this book will be of value to the neophyte as well as the experienced practitioners of electronic design and manufacturing, in particular, the small to medium-size companies that do not have the support staff and the resources necessary to try out some of the principles and techniques of green design and manufacturing and meld them into the company culture. The experiences documented here should be helpful to encourage many companies to venture out and develop new world-class green products that can make them grow and prosper for the future.

Sammy G. Shina

Acknowledgments

The principles of green design and manufacturing discussed in this book were learned, collected, and practiced through my 40 years in industry and academia. After graduation from the Massachusetts Institute of Technology, I worked in the electronics industry for 22 years, followed by another 20 years on the faculty of the University of Massachusetts Lowell. At the university, I worked as a teacher, researcher, and consultant to different companies, increasing my personal knowledge and experience in the fields of green electronic design, manufacturing, and quality.

I am indebted to several organizations for supporting and encouraging me during the lengthy time needed to collect my materials, write the chapters, and edit the book. I am thankful to the University of Massachusetts Lowell, for its continuing support for green design and manufacturing, especially the former provost, John Wooding, and the chairman of the department of mechanical engineering, John McKelliget. They both supported me in my research and work for the environment, and approved my application for a sabbatical semester to organize, edit, and write parts of this book. I am also appreciative of the Toxics Use Reduction Institute (TURI), especially its former and current directors, Ken Geiser and Michael Ellenbecker, as well as Liz Harriman and Greg Morose, who worked closely with me to manage both the Massachusetts and the New England Lead-Free Consortia. TURI has also supported my research and my students getting advanced engineering degrees in green materials and processes for more than 15 years, as early as the mid-1990s, starting with no-clean soldering, through additive PCB technologies, and into lead-free consortia.

My appreciation is also due to the New England Region of the EPA, who supported the consortium work in the latter years, and in particular Linda Darveau and Rob Guillimen, and my appreciation extends to them for awarding the Regional EPA 2006 Business, Industry and Professional Organizations Environmental Merit Award to the New England Lead-Free Consortium.

In addition, my thanks to Steve Chapman of McGraw-Hill who was my editor for this book, as well as my previous three books on Six Sigma and concurrent engineering. He always believed in me, encouraged and guided me through four books, and for that I am very grateful. I also wish to extend my gratitude to Vastavikta Sharma, Project Manager at ITC in Noida, India for her efficient and prompt editing and production of this book, as well as Toni Rafferty of Norah Head, NSW, Australia for hosting me during the editing and proofing of the book.

I am thankful to the many engineers who volunteered their valuable time and effort for the lead-free consortia that I created, and who participated through the many twists and turns of the electronics business, having started in one company and ending up in another. I wish them best success in implementing green products and methods in their companies.

Finally, many thanks to my family for emotional support during the writing, editing, and production of the book, including my wife, Jackie, and our children, Mike, Gail, Nancy, and Jon, as well as my grandchildren who brought me great joy between the many days of writing and editing.

Abbreviations

ANOVA	Analysis of variance
BSP	Board support package
CAD	Computer-aided design
CAE	Computer-aided engineering
CAF	Conductive anodic filament
CAM	Computer-aided manufacturing
CEM	Contract electronic manufacturer
C of C	Certificate of compliance
C_p	Capability of the process
df	Dissipation factor
DfM	Design for manufacture
DfE	Design for the environment
DIG	Direct immersion gold
DIP	Dual in position
D_k	Dielectric constant
DoE	Design of experiments
DOF	Degrees of freedom
EWERA	California's Electronic Waste Recycling Act
EMI	Electromagnetic interference
EMC	Electromagnetic Compatibility Directive
ENIG	Electroless nickel overplated with immersion gold
ENIPIG	Electroless nickel-immersion palladium-immersion gold
ERP	Enterprise requirements planning
ESD	Electrostatic discharge
ESI	Early supplier involvement
ESS	Environmental stress screening
EU	European Union
FMEA	Failure mode effects analysis
HASL	Hot air solder leveling
HASS	Highly accelerated stress screening
IEC	International Electrotechnical Commission
IMC	Intermetallic compound
IPC	Institute for Interconnecting and Packaging of Electronic Circuits

IC	Integrated circuit
ICT	In-circuit test
MBTF	Mean time between failures
MR	Moving range
MSL	Moisture sensitivity level
MSDS	Material Safety Data Sheets
NC	No-clean
NIH	Not invented here
N_x	Time to x percent failure
OA	Orthogonal array
OEM	Original equipment manufacturer
OSE	Odd-shaped eruption (in tin whiskers)
PCB	Printed-circuit board
ODM	Product data management
OSE	Odd-shaped eruption (tin whiskers)
OSP	Organic solderability preservative
PTH	Plated through hole
PPF	Preplated finish for ICs
PPM	Parts per million
QFD	Quality function deployment
QLF	Quality loss function
RoHS	Restriction of Hazardous Substances
RPD	Revolutionary product development
RSS	Root sum of the squares
RTV	Reliability test vehicle
SAM	Scanning acoustic microscopy
SIR	Surface insulation resistance
SMT	Surface mount technology
SL	Specification limit
SS	Sum of the squares
TCP	Trivalent chromium plating
T_d	Thermal decomposition temperature (for laminates)
T_g	Glass transition temperature (for laminates)
TH	Through hole (technology)
TQC	Total quality control
TQM	Total quality management
TURI	Toxics Use Reduction Institute
UBM	Under-bump metallization
VOC	Volatile organic compound
WEEE	Waste from Electrical and Electronic Equipment (EU directive)

CHAPTER 1

Environmental Progress in Electronics Products

1.1 Historical Perspective

The modern U.S. environmental movement began in the early 1960s with the Clean Air Act, which was originally passed in 1963 and extended in 1970. It progressed further with the advent of Earth Day in 1970, the Clean Water Act of 1972, and the establishment of the Environmental Protection Agency (EPA). During that time, many U.S. electronics companies were still in the integrated manufacturing mode, when they operated their own full-range facilities for printed-circuit board (PCB) fabrication and assembly as well as sheet metal and plating shops. The concepts of core competency and subcontracting had not yet taken hold. However, the hazards of certain metals and chemicals used in these facilities were beginning to be clearly recognized. There were major efforts at U.S. companies to chemically neutralize their chemical effluents and remove harmful compounds such as 1,1,1 trichloroethylene and cyanides from their processes. The EPA limits on the amounts of metals in the industrial effluents such as copper, lead, cadmium, mercury, and trivalent chromium were to be less than 1 part per million (ppm), and less than 0.1 ppm for hexavalent chromium. This is long before the current European Union (EU) restrictions on hazardous substances (RoHS) directives on lead, mercury, hexavalent chromium, polybrominated biphenyls (PBBs), and polybrominated diphenyl ethers (PBDEs) with proposed limits in products of 0.1 percent for all and with a limit of 0.01 percent for cadmium by June 30, 2006, thus effectively banning these materials.

The U.S. companies' reaction in the 1970s to the Clean Water Act was to initiate a crash program in cleaning up their industrial effluents. Many of the engineers drawn into this effort, like myself, were working in manufacturing either as electronics test engineers or in management positions. Consultants from the municipal waste areas were also brought in. The industrial effluent treatment consisted of changing processes to completely eliminate some of these hazardous materials, substitute other nonhazardous (green) materials, or chemically treat the industrial effluent generated by the hazardous materials and processes to precipitate the metal hydroxides. These collected hydroxides had to be compacted and then carted away to specially prepared landfills to contain them away from groundwater for eternity. The use and handling of hazardous materials by the production operators were regulated by controlling the environment through air handling and exhaust systems and were monitored by periodically testing the operators for the presence of harmful elements in their bodies. An example would be solder wave operators being tested for lead in their bloodstreams. The hazardous materials embedded in the products would be handled by recycling or scrapping the products in special landfills.

The RoHS approach to eliminating hazardous materials in the product is a much more effective method to create "green" products and processes. By eliminating the hazardous materials at the source, the need for extensive and special testing of operators, waste treatment facilities, and landfills would be removed, and the recycling programs could be made much simpler.

The initial efforts of companies to conform to the new environmental laws in the 1970s were somewhat rushed, leading to occasionally comical results, although not considered as such by the practitioners at that time. The engineers working on waste treatment of industrial effluents had to dust off their old freshman chemistry books to study chemical reactions, precipitations, and pH neutralization. The idea of 100 percent quality and keeping the industrial effluent neutral and metal-free 24/7 was alien to system design at that time. Most of the knowledge base came from the municipal water treatment technology and led to some bizarre suggestions, such as forming metal hydroxide precipitation lagoons at the end of the parking lots of companies. Most of the waste treatment facilities were built in a hurry, and not contained within new buildings to ensure proper operation in all weather conditions. One incident took place in a building owned by the now defunct Digital Equipment Corporation. A 3000-gal tank of lime (sodium hydroxide) was placed outside its manufacturing facility in Maynard, Massachusetts, to treat its industrial effluent. On a cold night, the tank and connecting pipes froze, leaked, and dumped the entire lime content into the Assabet River. The next morning it made the front-page news in the *Wall Street Journal*. This tragedy was receiving the reverse of what was hoped for: bad publicity from a good corporate deed.

The author was involved in a similar waste water treatment effort at the Hewlett-Packard Medical Division in Waltham, Massachusetts. The system received an award from the local water regulatory agency, the Metropolitan District Commission (MDC), for the design of the plant, which was featured on the front cover of its annual report in 1976. I had many memorable moments as a young engineer and a quick education in the risks and rewards of being an environmental change agent, as well as in the company and its suppliers' reactions to the first green efforts. They include the following:

- The elimination of 1,1,1 trichloroethylene and successful substitution of methylene chloride from a different supplier led to the original supplier accusing me, by formal letter to the management, of taking a bribe from the new supplier!

- When I explained to the operators who complained about the new materials not being as vigorous in cleaning parts as the old materials, that "methylene chloride is better for your health," they responded: "What about our health since we were using the old material for 20 years?"

- A near fatal accident occurred due to the elimination of cyanide from the conveyerized parts plating line. The line was reconfigured for noncyanide plating and recharged with a new acid solution, after a thorough cleaning. Unfortunately, traces of cyanide solution left over in the cracks of tanks in the line reacted with the new solution and spewed cyanide gas into the plating room. The room was evacuated very quickly with no injury to the operators.

- Eliminating hexavalent chromium and substituting trivalent chromium in the plating process required extensive testing with salt spray and adhesion testing of medical stethoscopes plated in the line. I still possess many of the functioning stethoscopes used in the salt spray test after 35 years.

- When some of the piping had to be reconfigured while the wastewater treatment plant was being turned on, there were pipe routing mistakes that had to be fixed promptly, or else they would have resulted in serious operational problems.

- The author had to complete a thorough EPA mandated yearly report listing all hazardous materials usage for the company. I had to obtain the signature of the division manager, Lewis Platt, who later became the CEO of the company. After examining the very thick document, he enquired as to what were the consequences of the report being inaccurate. I quibbled, "Someone has to go to jail!" and he responded, "That would be you!"

For today's green efforts in electronic products and processes, there is much better planning and a much longer time frame to develop good and reliable green material and process alternatives. For example, I started the Massachusetts Lead Free Consortium with local companies within the state in 1999, seven years before the RoHS deadline. That was plenty of time to research, examine, and test green material and process alternatives.

1.2   The Development of Green Design and Manufacturing Competency

One of the most difficult steps in the initiation of a green product development strategy is where to get started. The green knowledge base for product development is widely distributed and not readily available within the organization, in the design or process teams. The supply base has varying degrees of experience with green methods and processes. Several alternatives are available to bring this knowledge into the company: Groom the individual(s) within the organization, hire an experienced green professional away from another organization or competing company, or engage consulting services. Each alternative has its own benefits and drawbacks.

Internal green competency development takes time, and the choice of the individual(s) is very important. There are two elements to the green competency: One is the knowledge of the regulations, be they local, national, or international. The other is the knowledge of the process steps, the chemistry, and analysis tools required to evaluate the material properties for hazardous materials and their proposed green replacements. Several larger companies have opted for at least two required skill sets — one individual for regulatory and coordination with the appropriate agencies and another for material and chemical knowledge. The former could be an individual in the process, regulatory, or quality departments, while the latter could be a highly educated expert in materials as well as complex analysis methods. The majority of companies will probably choose one individual whose skills combine a mix of the two characteristics. However, this dual-skilled individual will be dependent on outside experts and their points of view when forming his or her opinions on green alternatives. That person might have difficulty wading through claims of competing trends in industry. In addition, he or she may have limited capability in undergoing extensive testing and analysis of green alternatives. In most cases where the company started with the two skill set model, one of the two will eventually leave the company or take on a different position internally.

Hiring away an expert from another organization or a competing company can benefit the company in the short term by quickly getting access to the individual's knowledge of green design and techniques. However, the company has to be cognizant of the individual's previous set of skills before she or he acquired the green knowledge and whether that is compatible with the company's future needs. As the green set of skills permeates the company's organization, the need for the green expertise of this individual will diminish, and the company might have to place this individual in a new position, where his or her previous skills could be useful for both the individual and the company.

Hiring a consultant or consulting services is very beneficial in two aspects: it is a quick method to obtain information and build the green competency, and it is also temporary, since the services are provided for a fixed time and there is no long-term commitment to the consultant. However, the drawbacks could be substantial and are listed below:

- The consultant might not be really up to date with green regulations or green process and material selection. A few questions could clarify the currency of the consultant's knowledge.

- The company should be wary of consultants recommending quick and easy green solutions: Do they have a personal stake in the materials or processes recommended? What is their association with the source of the recommended solution? Are there any hidden patents or copyright issues with their recommendations? Have they participated in studies sponsored and paid for by particular green suppliers? Have they declared all their prior or current associations with any supplier?

- Consultants deal with today's state of the art in green technology. In any new developing technology, the rate of innovation and research is high. Suppliers of green material and processes are constantly improving their products and can quickly leapfrog the performance of their competitors' materials. Thus the company will constantly have to go back to consulting services to keep ahead of technology or develop its own internal competency.

A mix of these strategies with short and long-term goals could be the best solution to the green competency challenge. The most important element is the planning stage, in order to devote enough time to formulate the appropriate green strategy and then form the green implementation team and key personnel to staff it.

1.3 Green Design and the New Product Life Cycle

In focusing on green design and manufacturing, it is important to understand the latest trends in the new product development cycle. The revolution in the high-technology industries has shrunk the product design and use life cycles to a period of weeks and months through concurrent engineering. At the same time, traditional design and manufacturing cycles in electronics circuits, tooling, and packaging had to be modified or outsourced to keep pace with new and lower-cost product introductions. The design team has been extended through the ubiquitous Internet to include collaborative activities within the company, its customers, and its suppliers.

The major premises of concurrent engineering have mostly been achieved, in terms of faster time to market, collocation of the various product creation team members to increase communications and feedback, and the use of design and quality metrics to monitor and improve the design process. The challenge is how to maintain and improve these gains and introduce green design and manufacturing at the same time by leveraging the trends in the globalization of design and manufacturing resources, and the wide use of the Internet as a communication tool.

The product realization process has undergone several changes with the advent of concurrent engineering. The change from a serial process of product development to a more parallel process has resulted in the need for new paradigms. Clearly, the impact of these new products is very critical, as indicated by vintage charts at different companies. In many high-technology companies, 70 percent of the total revenues of the company come from products introduced during the last few years.

Traditional product development required a top-down control of the various activities of product creation. Very formal organizational structures were developed and managed with a phase review process. Plans and milestones had to be completed at the end of each phase of product development and were subject to several levels of management reviews. After each review, the project was allowed to proceed and be funded until the next review.

The pressure toward shorter project time frames, global teams, quality and design, and manufacturing outsourcing have resulted in significant changes in the relationships between the company personnel and their suppliers with more frequent communications occurring earlier in the product development cycle. These suppliers and their own subsuppliers are called the *supply chain*.

1.3.1   Green Supply Chain Development

The major supply companies have mimicked the reach of the original equipment manufacturers (OEMs), by distributing their manufacturing centers globally, to be near their customers' sites. In this manner, supply companies can service global OEMs. The issues of global supply chain can be summed up as follows:

- The OEMs are forcing their suppliers to conform to their design specifications. For example, most OEMs will specify that the design and manufacturing documentation from suppliers conform exactly to their in-house CAE/CAD systems, including the system type and model number. In addition, OEMs can specify certain green materials and finishes and force the suppliers to produce them.

- The OEMs are also asking their suppliers for production quality verification, such as final testing, including troubleshooting of their products and systems. This could include reliability and quality testing of new green materials such as adhesion and pull tests of lead-free soldering by the suppliers, as well as certification of compliance with regulatory bans of hazardous materials (due diligence).

- Increased dependence on supplier quality and lower-cost goals has resulted in eliminating inspection for incoming parts by the company and shifting the burden to the suppliers, making companies vulnerable to spurious quality problems in the supply chain and increasing the need for due diligence on banned non green substances and materials.

- The trend toward increasing the links in the supply chain by further subcontracting to achieve even lower-cost manufacturing has resulted in low-technology suppliers getting into the manufacturing cycle for high-technology products. These suppliers do not have the sophisticated technology or the controls in place to make sure that all necessary specifications are inspected and variances in quality are promptly reported up the supply chain. For an OEM, a poorly managed supply chain is vulnerable to quality problems if changes are made in the subcontractor chain without the OEM's approval or notification. It is recommended that the supply chain not extend beyond three levels down from the final assembly.

- These low-technology suppliers represent a greater risk to the green supply chain. They tend to be unregulated, with greater focus on cost than on quality or green. They might purposely substitute specified green materials, which might be more expensive, with nongreen components or processes. A very effective system of quality and due diligence has to be in place to prevent this tendency. Recent examples of pet food and toy finish poisoning from China show the large negative impact of unregulated foreign suppliers.

- The electronics supply chain is also performing warranty and repair of the customer products. For green products and processes, the repair process of electronic circuits has to be developed and tested, since these processes might have different characteristics from nongreen ones. Other chapters in this book explore how to develop and test repair processes for electronic circuits and PCBs.

1.4   Adverse Design Consequences to Improper Adoption of Green Materials and Processes

The obvious concerns in adopting green materials and processes are the quality and reliability issues. In this book there are several chapters with many examples of case studies on how to properly test green materials and processes for quality and reliability, using either mathematical or physical environmental conditions to simulate life-cycle material behavior. These tests can be performed by either comparing the properties of green materials and processes to a baseline of nongreen materials, or to study the life cycle through testing to failures. These techniques will result in calculating a quantifiable risk level for green alternatives, or lack thereof.

Most of the green materials and process verifications in this book and in the general literature are based on the performance of the green materials, and not on the consequences of adopting new green materials on the functional performance of the products. New products can be based on either existing designs that have been converted to green or an improvement on the technology of an earlier generation of nongreen products. In the case of converting current products to green, some product performance verification tests will have to be repeated to ensure compliance with advertised specifications, especially when PCB designs are re-layed because of lack of green replacement components with the same footprints. In the case of newly designed green products, verifications tests will uncover any hidden problem when using fresh green materials and processes in the design.

While a company might perform a very thorough set of tests and analysis of green materials and processes, the green products using them might exhibit different functionality than expected in the nongreen predecessor products. Refer to Sec. 7.2 for more insight into the green material conversion consequences in design. These potential adverse consequences could be as follows:

1. Radio-frequency (RF) and wireless circuits. These circuits are susceptible to changes in the shielding provided by changes in the metallic or plating of shield metals (such as the removal of hexavalent chrome coating) or by changing of the PCB surface finish [from tin lead hot air solder leveling (HASL) to immersion silver, for example]. These new green materials might cause a frequency shift in the tuners of these circuits. This is especially serious if there are no adjustments available to tune the RF or wireless oscillators. Redesign or relayout of the electronics circuits might be required.

2. Clock speeds and propagation delays. In very high speed circuits, clock speeds and propagation delays through the electronics circuits and components might be affected by

changes to the surface finish of the PCBs or to the materials used in the interconnecting traces on the outer layer surface and through the inner layers of multilayer PCBs and connectors. Electronic race conditions might also be increased due to these conditions.

3. Transmission line reflections. When the electronics on the PCBs are connected directly to outside connectors or to other PCBs or products in a digital communications mode, the impedance match between them is very critical to sustain digital transmission over specified connector cable lengths. Changes to green materials might offset this impedance balance, rendering communications less effective. The communications system might work for shorter distances than specified, engendering a serious customer satisfaction issue. A simple PCB test would be the standard Time Domain Reflectrometry (TDR) test, where a coupon with a specified parallel trace length is used for the PCB laminate to ensure transmission line conformance to specification. This test can be performed at the PCB fabricator location and specified as part of the quality audit.

4. PCB surface features: In some electronic designs, PCB surface features are used for electrical, magnetic, or charge coupling in the electronics circuits. Examples would be the use of filled-in surface pads or rings to generate proportional feedback for a position control circuit. Changing the PCB surface finish to green materials might change the electrical or electromagnetic properties, rendering the circuit inoperable or out of control.

5. Surface resistance and cleanliness: With new green materials for soldering and fluxing of electronic PCBs, the cleanliness properties of the PCB after soldering or reflow might change. These changes can be easily measured with a variety of instruments such as surface insulation resistance (SIR) meters.

1.5 Implementing a Successful Green Design and Manufacturing of Products, Using Quality Tools and Techniques

The conversion of electronics manufacturing processes to comply with RoHS regulations provides a great opportunity of process improvement for efficient as well as environmentally friendly design and manufacturing processes. Six Sigma and quality techniques for process measurement, analysis, and improvement can be used to select optimum RoHS-compliant processes that will increase quality and reduce cost. The methodologies outlined in this section were used by the New England Lead Free Consortium of companies created by the author and jointly supported by funding and resources from member companies, the Toxics Use Reduction Institute (TURI), and the EPA.

An unintended consequence of the RoHS directive began when the components suppliers implemented their RoHS compliance by eliminating banned substances such as lead in the component finishes. As a result, some suppliers decided not offer the components with the original lead finishes, since they did not want to keep two versions of the same component. The exempted industries are finding that they cannot easily obtain their traditional leaded components and therefore are being forced to make necessary changes to their components and processes to be RoHS-compliant. It appears that an unintended impact of the RoHS regulation is a universal switch away from the banned substances for all industries, giving all companies the opportunity to make optimum material and process improvements as they switch into new materials and processes because of RoHS compliance.

Implementation of RoHS offers several opportunities as well as dilemmas for companies. This is because of the myriad alternatives being proposed by their suppliers with conflicting claims, as well as the pace of technological progress in material technologies. What is a hot RoHS material substitute today might become out of favor because of subsequent developments. Examples include such developments as bismuth-based solders, which were attractive because of their lower melting temperature that have fallen out of favor since they have contamination problems with lead, while tin-based solders and finishes were suspect as lead-free replacements because of their higher reflow temperatures and the dreaded tin whiskers. In addition, technological developments in material technology, as in any other, tend to produce leapfrog effects: a supplier that claims to have the best results for their materials might be overtaken by another supplier with a newer technology. So what is a company to do that is trying to implement RoHS?

The larger companies developed their own research program, with a multitude of talent and resources brought to bear, to solve the problems of material and process conversion for RoHS compliance. They might work with their contractors and material suppliers to implement RoHS compliance using a variety of reliability and test methods, as well as complex analytical tools such as vibration platforms, long-term environmental testing to failure, and electronic scanning microscopes to see the interfacing layers of RoHS-compliant materials. These sophisticated analysis tools and extensive DoE matrices were used to evaluate a large number of candidate material replacement and process parameters.

Medium and smaller companies could not afford these massive programs, but could develop cost-effective green conversion programs using Six Sigma principles for RoHS conversion projects. A set of simple guidelines to effectively manage a successful conversion could be as follows:

1. Avoid the NIH (not invented here) syndrome. There are many resources available for identifying successful materials and process replacements for RoHS-prohibited materials. These include national consortia and standards-setting organizations for electronic products such as NEMI, SMTA, and IPC. However, these organizations might recommend the general composition of the replacement materials, but not the specific process parameters to handle the local complement of equipment and product mix that the company uses.

2. Use standard performance criteria and test methods for the RoHS materials whenever possible. Standards for reliability testing using temperature and humidity cycling, vibrations, electrical conductivity, and mechanical stress exist for many materials from the same sources mentioned in the preceding paragraph. In addition, use testing methods that are standardized in industry, either by using techniques that are outlined in the standards or using commonly available commercial testers to perform the test.

3. When standardized or commonly used testing methods are not available, the current processes could be used as the baseline when comparing RoHS-compliant materials to the current process. Comparing to the baseline can also be used when there is not enough time or resources to properly conduct the testing for the RoHS materials. This could be the case for shorter-term environmental testing. Examples would be to use fewer temperature cycles and not test to failure, but to compare pull tests of RoHS materials to baseline leaded counterparts. Vibration testing could be on specific spectrum lines and not the full spectrum of frequencies. Statistical significance testing could be used to compare lead-free results to the current product leaded manufacturing process baselines.

4. Use the latest green material selection available, from leading material suppliers, realizing that today's RoHS material champion might not be the champion tomorrow, and that the tests might have to be repeated down the road as the material technology keeps improving. Avoid proprietary or patented materials.

5. Realize that the RoHS conversion effort might not be performed in one single large continuum project, but might be a succession of smaller projects that builds on the knowledge acquired in the previous project. For example, in lead-free soldering, there might be an initial project to deal with SMT

components, with a distinct portion for BGA/mini BGAs, and then follow on projects for through hole and rework.

Research efforts for RoHS conversion by the New England Lead Free Consortium of companies and academia, which was created by the author, began in 1999 and continue today. The research followed the broad guidelines outlined above. The concerns for RoHS conversion were focused on three parts: reliability, quality, and manufacturability. The RoHS-compliant alternatives had to meet and/or exceed current nongreen materials and processes in terms of reliability; they should be able to produce virtually defect-free products and be implemented in a typical manufacturing process line with standard machines and processes, including repair and rework.

The projects conducted for the lead-free conversion are comprised of four major phases to date, with each phase taking about 2 years. The lengthy time is due to the consensus needed to be achieved by the consortium members, and the funding requirements, as most materials and actual manufacture and tests were donated or performed by member companies, on a voluntary and pro bono basis. More information, including papers published on these projects, could be obtained from the author or the TURI website at www.turi. org. More detailed information on these phases is available in the next chapters, showing the technical background and reasoning behind the decisions made. The phases were as follows:

1. *Feasibility (phase 1)*. During this phase, the feasibility of lead-free soldering was explored as a viable alternative to leaded solder. The goal of this phase was to provide knowledge on the major issues of lead-free implementation during that time frame (1999 to 2001): What solder composition is the best alternative to tin/lead [(tin/bismuth, tin/silver/copper (SAC), and tin/silver]? What about the higher melting temperatures? Can the thermal energy to melt the solder be integrated in time to lessen the impact of the thermal shock on electronic components? Can a lead-free process deliver zero defects under controlled conditions?

   Table 2.12 details the design of experiments (DoE) matrix for phase 1, showing the selection of 5 factors in 27 experiments. TAL refers to time above liquidus. The project material and process selection was limited to a small number of laminate and component finishes. The selection was also organized in a set of partial factorial experiments to lessen the time and effort involved. The reliability testing was performed through pull tests after 2000 typical thermal cycles of 0 to 100°C in 1-h cycles, and the number of solder defects was analyzed on a ppm (parts per million) basis, according to IPC standards.

   The results were very encouraging. Reliability data (thermal cycling followed by pull tests) showed that the lead-free joints were stronger than legacy tin lead joints, and the quality data indicated that zero defects were possible with lead-free soldering in certain combinations of solders and PCB surface finishes. Manufacturability issues of thermal profiling were also shown to be of little or no significance.

2. *Wide material selection (phase 2).* In this phase, the experience gained from the first phase was used to narrow some of the choices such as the SAC solder formulation and a common thermal reflow profile, while the alternatives for laminate finish (5) and component type surface finishes (4) were expanded. Manufacturability concerns such as the use of nitrogen were also added. A full factorial experiment was performed to make sure that there was no confounding of interactions. A baseline of legacy tin lead finish components was also produced and compared to the lead-free alternatives. This experiment was much larger in scope and effort than the first experiment. The same reliability assessment of thermal cycling and pull tests as well as quality consideration of 100 percent visual testing according to IPC standards was used, in addition to some investigations of the inner metallic layers using SEM.

The results were very similar to those of phase 1, in that certain combinations of materials and processes produced near zero solder defects. Many of the factors were not significant in either reliability or quality testing. Figure 1.1 is a Minitab® plot that shows the distribution of visual defects versus the selected factors of surface finish, solder suppliers, and reflow atmosphere. Subsequent analyses showed the selections of the finishes were not significant to one another in quality or reliability data.

An interesting subtest of the analysis was that under certain combinations of materials and processes, the quality and reliability

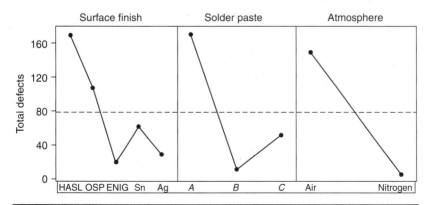

Figure 1.1 Phase 2 Graphical analysis of Visual defects.

of lead-free soldering were not affected by whether nitrogen was used in the reflow process. This is an important finding in terms of reducing the cost of electronics manufacturing.

3. *Manufacturing process optimization* (phase 3, also called TURI TV3 in Chap. 6). Building on the previous phase, the material and process selection was narrowed down to the successful candidates gleaned from phase 2, and then the project focused on simulating actual manufacturing conditions. A larger test PCB was manufactured, at the standard panel size of 16×18 inches as opposed to the 6×9 inches from phase 2.

Although a full factorial experiment was conducted, the total number of test PCBs that were analyzed was at 24 lead-free PCBs as well as 12 for a leaded PCB baseline, compared to 120 PCBs from phase 3. The testing performed was similar to phase 2, with the addition of vibrations and multiple reflows (to simulate rework and repair) to the mix. The reliability and quality results for surface mount technologies (SMTs) indicated that with the factor selection used, there were no significant differences between the lead-free and the leaded baseline PCBs in reliability as measured by the pull tests, and in quality, as measured by visual inspection. The significant differences in pull tests were in three areas: lead finish (tin versus tin-bismuth), pull direction (up or down, because of the location of the pull gage), and the interaction of laminate finish and laminate type (one of the two SAC solder suppliers performed significantly different with the three laminate types shown). Figure 1.2 is a plot generated by

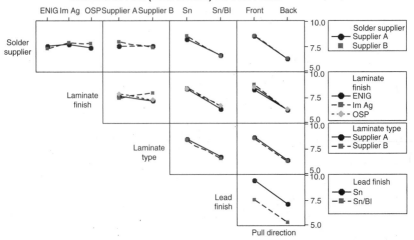

FIGURE 1.2   Interaction plots of pull test of lead-free project Phase 3.

Minitab® to show the various interactions present in the DoE of phase 3. Quality analysis showed a significant lower quality for through hole and rework conditions, necessitating a closer look at these conditions in phase 4, and recounted in Chap. 6 in this book.

4. *Manufacturing and technology optimization extension* (phase 4, also called TURI TV4 in Chap. 6). In this phase, the unresolved issues of phase 3 were further investigated. These include the through hole technology and the rework methodology. In addition, new materials that might be included in future green efforts such as halogen-free laminates are being investigated against a baseline of halogen-based materials.

Similar efforts were undertaken to comply with RoHS requirements for other prohibited materials, including hexavalent chromium, done in the University of Massachusetts Lowell in association with Tyco Electronics. Several alternatives were examined, and a replacement chart for each application was generated based on extensive testing in harsh environment, all meeting company and industry standards. This successfully concluded project followed the general guidelines given above.

In conclusion, the RoHS conversion process for prohibited materials is an excellent opportunity to examine the material replacements and their manufacturing processes and to apply Six Sigma and quality principles to improve quality and lower manufacturing costs.

1.6   Organization of the Rest of the Chapters for *Green Electronics Design and Manufacturing*

The chapters in this book and their authors were selected for their specific expertise in different green competencies outlined in this chapter and represent the whole spectrum of green product design, development, manufacturing processes, and the electronics supply chain. Their dedicated work, supported by their team members and their companies, represents what is best about the modern global electronics industry: a true belief in hard work to develop green products and processes while at the same time maintaining the companies' responsibilities to their employees, customers, and shareholders. They represent what is optimistic about the global electronics industry: once members of the industry realized that they could do a better job in creating green products and processes, they went ahead, rolled up their sleeves, and went about solving the challenges of green design and manufacturing. This book will document their successful efforts, by distilling their knowledge and experience and presenting their companies' success stories. The book is divided into three parts.

Part I

Part I is the scientific and mathematical underpinning to successful conversion of materials, products, and processes to green, represented in two chapters on quality and reliability.

Chapter 2: Statistical Analysis of Green Electronic Products

By Sammy Shina, University of Massachusetts Lowell, sammy_shina@uml.edu

This chapter covers the following: Principles of statistics and design of experiments (DoE). Types of statistical tests as applied to lead-free and green material and process selection. Determining process characteristics, for the short and long terms. Two- and three-level experiments. Graphical and statistical analysis of lead-free experiments' material and process selection. Full and partial factorial lead-free experiments and their interactions.

Chapter 3: Reliability of Green Electronic Systems

By David Pinsky of Raytheon Company, david_a_pinsky@raytheon.com

This chapter covers the following: Principles of electronic product and component reliability. Types of reliability evaluations and analysis: thermal and mechanical reliability testing (thermal cycling strategy types and methodologies, vibrations, drop, HALT). Unintended consequences of green product conversions (new types of defects, tin whiskers).

Part II

Part II covers the planning and implementing of green strategy and products at the corporate strategy and project management level.

Chapter 4: Environmental Compliance Strategy and Integration

By Ken Degan, Teradyne Corporation, Ken.degan@teradyne.com

This chapter covers the following: Developing a green strategy ahead of the new product development cycle. Keeping up with environmental knowledge, trends, issues, and regulations. Guiding the design teams for developing green products. Understanding the green impact on the organization, design methods, and techniques.

Chapter 5: Managing the Global Design Team in Compliance with Green Design and Manufacturing

By Mark Quealy, Schneider Automation, Mark.Quealy@us.schneider-electric.com

This chapter covers the following: Planning green-compliant new product design. Global compliance issues. Decision making and lessons learned form converting existing products and processes to green. Local and global teams skills distribution. Data and documentation processes required.

Part III

Part III is a collection of topics of green product design and process development and conversion from the component, manufacturing, and the supply chain, now and into the future. They include contract electronics manufacturing at Benchmark Electronics, in-house manufacturing at Tyco Electronics, laminate and PCB fabrication at DDi Incorporated, electronics components at Texas Instrument, and lead-free nanosolder as well as other electronics materials development at NanoDynamics Inc.

Chapter 6: Successful Conversion to Lead-Free Assembly

By Robert Farrell, Robert.farrell@bench.com, and Scott Mazur, Scott.Mazur@bench.com, of Benchmark Corporation

This chapter covers the following: Planning for conversion of current assembly operations from legacy to lead-free operations. The component, material, and equipment decision process required for successful lead-free conversion. Managing the mix of lead and lead-free component finishes, including sourcing and alternate processing methods. Planning for lead-free assembly process modifications. Training of the assembly associates and inspectors. Rework and repair issues of lead-free products.

Chapter 7: Establishing a Master Plan for Implementing the Use of Green Materials and Processes in New Products

By Dick Anderson, of Tyco Electronics, andersonr@tycoelectronics.com

This chapter covers the following: Working with the design teams to plan for green materials, manufacturing, cables, and packaging. Establishing a master plan for implementing the use of green materials in new products. Product conversion to lead-free products. Conversion of new products to Hex-Chrome-free products. Green cabling planning and implementation.

Chapter 8: Evolution of Green Laminates and Fabricated Printed Wiring Boards, Types, Properties, and Materials Properties

By Mike Taylor, Dynamics Details Incorporated, Longmont, Colorado, mtaylor@ddiglobal.com

This chapter covers the following: Laminates types and classifications. Selection process for green and halogen-free laminates. Compatibility of laminates to green and lead-free solders and component finishes. Safety and green requirements for laminate materials. Cost impact of green and halogen-free laminates.

Chapter 9: Green Finishes for IC Components

By Donald Abbott, Texas Instruments, Attleboro, Massachusetts, d-abbott1@ti.com

This chapter covers the following: Historical evolution of IC finishes. Types of IC leadframe geometry and packaging methodologies. Types of IC finishes and their green attributes. Adverse consequences of IC finishes: tin whiskers and moisture sensitivity levels.

Chapter 10: Nanotechnology Opportunities in Green Electronics
By Alan Rae, Ph.D., Nanodynamics, arae@nanodynamics.com
This chapter covers the following: The concepts of nanomaterials and nanotechnologies and their use in electronics products. Nanotrends in solders and die attachment. Examples of current application of nanotechnology in electronic products.

CHAPTER 2

Statistical Analysis of Green Electronic Products

In experimenting with green materials and processes, the most common methodology is to compare and contrast the new green material or process properties with current nongreen materials and processes. The usual concern is the necessary volume of samples taken for the green materials or process, as compared to current production material and process profiles, usually expressed as the capability of the process (Cp) or process capability.

There are two parts to this concern: The immediate reaction is that a very large volume of green samples has to be made to properly compare to nongreen samples. The other part is that the measurements used in current nongreen material and production processes for quality control and defect rate prediction might not apply because of the difficulty of properly obtaining statistical information such as the standard deviation of the manufacturing processes. Low-volume industries including defense, aerospace, and medical, as well as their suppliers, are concerned because they do not have the current production volume to justify large test samples of green alternatives to current materials.

Several statistical tools are discussed in this chapter to allow for the use of statistics in low-volume production environments, with minimum uncertainties. The tools are based on sampling theory and distribution as well as the relationships between samples and populations. In addition, the techniques of design of experiments (DoE) are discussed in order to select the best performing materials or processes from multiple green alternatives.

Section 2.1 discusses the sample probability distribution and its relationship to the parent population distribution. It gives an example of determining the population standard deviation and error based on sample sizes.

Section 2.2 discusses the amount of data required to properly determine process capability of materials and processes. The data volume is important in increasing the accuracy and amount of effort necessary to make sure that green materials and processes have equivalent properties to the current nongreen materials. This section also presents the moving-range method as a source of measuring process capability in low volumes.

Section 2.3 discusses the issues of determining process capability during the different stages of the green product life cycle. The strategies of setting different quality expectations during prototype versus volume production are also examined.

Section 2.4 discusses the methodology used to investigate potential green substitutes for nongreen materials and processes. Different types of DoEs are reviewed and recommended for different phases of the green selection process, together with the interpretation of results and explanation of the significance and confidence in the recommended solution.

2.1   Process Average and Standard Deviation Calculations for Samples and Population

The knowledge of certain properties of a subset, or sample, can be used to draw conclusions about the properties of the whole set, or population. Properties can be of two types:

1. *Quantitative (variable):* These properties can be observed and recorded in units of measure such as the thickness of deposition. The units should be produced under the same replicating conditions.

2. *Qualitative (attribute):* These properties can be observed when units are being tested under the same set of gages or test equipment. The result is either an acceptable (good) part that meets specifications or a rejected (bad) part that does not. An example is the set of all thicknesses of materials produced in the same conditions and measured against a known upper or lower limit using a go/no-go gage.

An important distinction between samples and populations is the sample size n: In the random choice of n objects from a population, each is independent of the others. As n approaches infinity, the sample distribution values of average and standard deviation become equal to those of the population.

Variable control charts, most commonly used in production quality, constitute a distribution of sample averages, with constant sample size n. This distribution is always normal, even if the parent population distribution is not normal. It has also been shown that the standard deviation s of the distribution of sample averages is related to the parent population distribution standard deviation σ by the central limit theorem, which states that $s = \sigma/\sqrt{n}$. The number of samples needed to construct the variable chart control limits is usually set at a high level of 30 successive samples to ensure that the population σ will be known.

When the total number in the samples n is small, very little can be determined by the sampling distribution for small values of n, unless an assumption is made that the sample comes from a normal distribution. In the normal distribution, an infinite amount of occurrences are represented by the population process average μ and standard deviation σ. The Student's t distribution is used when n is small. It does not require knowledge of the population standard deviation σ. The data needed to construct this distribution are the sample average $\overline{X}$ and sample standard deviation s as well as the parent normal distribution average μ.

$$t = \frac{\overline{X} - \mu}{s / \sqrt{n}} \qquad (2.1)$$

where t is a random variable having the t distribution with $v = n - 1$

$$v = \text{degrees of freedom (DOF)} = n - 1 \qquad (2.2)$$

The shape of the t distribution is similar to that of the normal distribution. Both are bell-shaped and distributed symmetrically about the average. The t distribution average is equal to zero, and the number of degrees of freedom governs each t distribution. The spread of the distribution decreases as the number of degrees of freedom increases. The variance of the t distribution always exceeds 1, but it approaches 1 when the number n approaches infinity. At that time, the t distribution becomes equal to the normal distribution.

The t distribution can be used to determine the area under the curve, called the significance, or α, versus a given t value. However, the t distribution is different from the normal distribution in that the number in the sample and degrees of freedom v have to be considered. The table output value of variable t, called t_α, is given, corresponding to each area under the t distribution curve to the right of α and with v degrees of freedom. The term *significance* is not commonly used, but its complement is called the *confidence*, which is set to equal 1 − significance and is expressed as a percent value.

$$\text{Confidence (\%)} = 1 - \text{significance} = 1 - \alpha \qquad (2.3)$$

Table 2.1 is a selected set of the values of t_α. The t distribution is used in statistics to confirm or refute a particular claim about a sample versus the population average. It is always assumed that the parent distribution of the t distribution is normal. This is not easily verified

v	$\alpha = 0.10$	$\alpha = 0.05$	$\alpha = 0.025$	$\alpha = 0.01$	$\alpha = 0.005$	$\alpha = 0.001$	$\alpha = 0.0005$
1	3.078	6.314	12.706	31.821	63.657	318.3	636.6
2	1.886	2.920	4.303	6.965	9.925	22.327	31.600
3	1.638	2.353	3.182	4.541	5.841	10.214	12.922
4	1.533	2.132	2.776	3.747	4.604	7.173	8.610
5	1.476	2.015	2.571	3.365	4.032	5.893	6.869
6	1.440	1.943	2.447	3.143	3.707	5.208	5.959
7	1.415	1.895	2.365	2.998	3.499	4.785	5.408
8	1.397	1.860	2.306	2.896	3.355	4.501	5.041
9	1.383	1.833	2.262	2.821	3.250	4.297	4.781
10	1.372	1.812	2.228	2.764	3.169	4.144	4.587
20	1.325	1.725	2.086	2.528	2.845	3.552	3.849
30	1.310	1.697	2.042	2.457	2.750	3.386	3.646
∞ ($t = z$)	1.282	1.645	1.960	2.326	2.576	3.090	3.291
Confidence or $1 - \alpha$	90%	95%	97.5%	99%	99.95%	99.9%	99.995%

TABLE 2.1 Selected Values of $t_{\alpha,v}$ of Student's t Distribution

since the sample size is small. There are several methods to ascertain whether the data are normal: plotting the sample data on normal paper which can show approximately whether the data distribution is normal if the plot is linear, or plotting a normalized score of the data and observing linearity, or using one of the many statistical software packages available.

Historically, the confidence percent used depended on the particular products being made. For commercial products, a 95 percent confidence level is sufficient, while for medical and defense products where higher reliability is needed, 99 percent confidence levels have been used. When the confidence percent increases, the span of the confidence interval also increases as well as its endpoints, the confidence limits. For low-volume data, the confidence limits for the population average μ and standard deviation σ estimates are used to give an approximation of the span of these two variables. The 95 percent confidence limits can be used for calculating the confidence interval (the $\pm$ range of the error in the number), while higher confidence numbers (99 and 99.9 percent) can be used as worst-case conditions to check on the base calculations. An example of using the t distribution in comparing tin lead and lead-free quality as indicated by visual inspection is given in Tables 2.3 and 2.4 and explained later in the chapter.

2.1.1   Other Statistical Tools: Error, Sample Size, and Point and Interval Estimation

Section 2.1 introduced statistical terms that are not in wide use by engineers, but are very familiar to statisticians. This section is a review of some of the statistical terms and procedures dealing with error estimation for the average and standard deviation as well as their confidence limits.

A good number to use for statistically significant data is 30. It is a good threshold when one is using some of the quality processes such as calculating defect rates. This is based on the fact that a t distribution with ν degrees of freedom = 29 approaches the normal distribution. The data for the value of $t_{\alpha,30}$ are close to the value of the standard normal distribution. The error E is calculated as the difference between the $t_{\alpha,30}$ value and the z value from the normal distribution. For a significance of 0.025 from Table 2.1, or confidence of 97.5 percent, the error is less than 5 percent. Note that this point of $z = 1.96$ is close to the $z = 2$ or the 2σ point or 95 percent confidence span. For the 3σ point, or 99.9 percent span, the error approaches 10 percent. The conclusions of the t test can thus be determined with small samples and with known errors.

The relationship between the error and the sample size can be expanded to include the general conditions in which the standard deviation σ is known from the sample and the numbers in the sample taken are large (>30). The maximum error E produced when a particular sample size n is used can be calculated in the following equation. Conversely, the random sample size needed to estimate the average of a population, with a confidence of $(1 - \alpha)$ percent can also be shown as

$$E = z_{\alpha/2} \cdot \sigma / \sqrt{n}$$

and   (2.4)

$$n = \left(\frac{z_{\alpha/2} \cdot \sigma}{E} \right)^2$$

where E is the error, σ is the standard deviations of the population, and n is the sample size used in calculating the error. The two equations in (2.4) show how to derive the error from the sample size taken if the standard deviation σ is known and, conversely, how large a sample is to be taken if a certain level of error is desired.

If the sample size n is small (<20) and the sample is drawn from a normal distribution of the population, the standard deviation of population σ is not known, but the sample standard deviation s can be calculated from the sample. In this case, the error made when the sample average $\overline{X}$ is used to estimate the population average μ is as follows:

$$E = t_{\alpha/2} \cdot \frac{\sigma}{\sqrt{n}}$$ (2.5)

2.1.2   Confidence Interval Estimation for the Average

Engineers have found the use of confidence percent in Sec. 2.1.1 for estimating the average rather unfamiliar. They are more comfortable with the concept of the confidence interval. This term shows the range of the average having the degree of confidence $(1 - \alpha)$ percent. The endpoints are referred to as the confidence limits. The formulas for the interval of the average estimation are for high- and low-volume samples, respectively:

$$\overline{X} - z_{\alpha/2} \cdot \frac{\sigma}{\sqrt{n}} < \mu < \overline{X} + z_{\alpha/2} \cdot \frac{\sigma}{\sqrt{n}}$$ (2.6)

and

$$\overline{X} - t_{\alpha/2} \cdot \frac{s}{\sqrt{n}} < \mu < \overline{X} + t_{\alpha/2} \cdot \frac{s}{\sqrt{n}}$$ (2.7)

If the confidence limit was at 95 percent (or $z = 2\sigma$ away from the average), then it is expected that the probability of at least one interval span falling outside the population average is 5 percent, or 1 out of 20 samples. Therefore, a sample the average interval span of which does not contain the population average is considered unlikely to happen.

2.1.3   Standard Deviation for Samples and Populations

The statistical relationships of the sample and population averages have been discussed in previous sections. There is a similar distribution for the sample variability S^2, which can be used to learn about its parametric counterpart, the population variance σ^2. This distribution is called the *chi squared* or χ^2. Since the distribution cannot be negative, it is not symmetrical, but in fact is related to the gamma distribution. The probability that a random sample produces a χ^2 greater than some specified value is equal to the area of the curve to the right of the value. The variable χ^2_α represents the value of χ^2 above which there is the area α. The equation for the distribution variable is

$$\chi^2 = \frac{(n-1)\dot{s}^2}{\sigma^2} \qquad (2.8)$$

where s^2 is the variance of a random sample of size n taken from a normal population having the variance σ^2 and χ^2 is a random variable having the distribution with degrees of freedom $v = n - 1$.

Table 2.2 contains selected values of the χ^2 distribution. Since it is not symmetrical, two χ^2 values will have to be returned when confidence percents are needed for two-sided limits. As in the t distribution, the χ^2 distribution can be used in two cases:

- When the population variance σ^2 is known, and therefore the probability that the sample variance s^2 can be tested to see if it is related to the population variance σ^2

- When the population variance σ^2 is not known, and the sample variance s^2 is used to determine σ^2, with confidence limits and confidence intervals. The equation for this case is

$$\frac{(n-1)s^2}{\chi^2_{\alpha/2}} < \sigma^2 < \frac{(n-1)s^2}{\chi^2_{1-\alpha/2}} \qquad (2.9)$$

where s^2 is the variance of a random sample of size n from a normal population, the confidence interval for σ^2 is $(1 - \alpha)$ percent, and $\chi^2_{1-\alpha/2}$ and $\chi^2_{\alpha/2}$ are values having areas of $1 - \alpha/2$ and $\alpha/2$ to the right and left, respectively, of the distribution average.

ν	$\alpha = 0.995$	0.975	0.95	0.90	0.50	0.10	0.05	0.025	0.005
1	0.0000393	0.000982	0.00393	0.0158	0.455	2.706	3.841	5.024	7.879
2	0.0100	0.0506	0.103	0.211	1.386	4.605	5.991	7.378	10.597
3	0.0717	0.216	0.352	0.584	2.366	6.251	7.815	9.348	12.838
4	0.207	0.484	0.711	1.064	3.357	7.779	9.488	11.143	14.860
5	0.412	0.831	1.145	1.610	4.351	9.236	11.070	12.832	16.750
6	0.676	1.237	1.635	2.204	5.348	10.645	12.592	14.449	18.548
7	0.989	1.690	2.167	2.833	6.346	12.017	14.067	16.013	20.278
8	1.344	2.180	2.733	3.490	7.344	13.362	15.507	17.535	21.955
9	1.735	2.700	3.325	4.168	8.343	14.684	16.919	19.023	23.589
10	2.156	3.247	3.940	4.865	9.342	15.987	18.307	20.483	25.188
15	4.601	6.26292	7.261	8.547	14.339	22.307	24.996	27.488	32.801
20	7.434	9.591	10.851	12.443	19.337	28.412	31.410	34.170	39.997
25	10.520	13.120	14.611	16.473	24.337	34.382	37.652	40.646	46.928
30	13.787	16.791	18.493	20.559	29.336	40.256	43.773	46.979	53.672

TABLE 2.2 Selected Values of χ^2 Distribution

2.2   Determining Process Capability

Process capability is the analysis of a process to determine its quality. A single quality characteristic or several quality characteristics are selected, some of which might be variable or attribute. For variable characteristics, the distribution of the data collected is checked for normality, and the distribution average μ and standard deviation σ are calculated. It has been shown in this chapter that it takes a sample size of 30 measurements to directly obtain these two parameters and determine whether the distribution of data is normal. For low-volume production, Sec. 2.1 discussed methods of determining a confidence interval for the two parameters. The confidence limits from these intervals could be used for worst-case determination of quality. For attribute processes, the defect rate is determined for parts that are manufactured in small quantities as prototypes, or from similar parts in current production.

The amount of sampling required for determining process capability is also dependent on whether the process has been in production (existing) for some time or whether a new process is being created. Once the process is operating on a regular basis and a reasonable level of quality is achieved, it is also desirable that the quality characteristic(s) being measured be charted for statistical control in control charts. For quality level approaching Six Sigma and beyond, control charting might not be required, and can be substituted with a total quality management program to monitor individual defects per period as opposed to using the sampling methods of control charts.

2.2.1   Process Capability for Large-Volume Production

The following procedures are recommended when time and resources are not a gating item. They are ideally suited for large-volume manufacturing where the parts cost is low and the ease of collecting data is high. These procedures will increase the accuracy of the process capability and reduce its apparent variation with time.

1. *Initial determination of process capability:* Historical guidelines for variable and attribute data are based on many subgroups of data, each at 30 samples. Each subgroup of data should be taken at different times, preferably on different days. In this manner, day-to-day variations of the process could be integrated into the process capability calculations. There should be no allowance for process average shift in the C_{pk} calculations.

   For low-volume applications, the moving-range method should be used because of the low volume required. A discussion of the moving-range method is given in Sec. 2.2.2.

2. *Regular updates of the process capability:* The process capability should be regularly checked to determine if the process has

changed. If the change is deemed significant using statistical tests, then a process quality correction project should be initiated to determine the cause of the process deviation.

The amount of data points required for checking the process could be less than the original data needed for initial determination. Determination of σ can be achieved either directly from the data or through $\bar{R}$ the estimator for variable data. For large-volume production, a sample size of 30 is sufficient to perform this check of process capability for variable data. For low-volume production, smaller sample sizes can be used and deviations tested for the probability that the average or standard deviation has shifted from the original, given a confidence interval.

3. *Correction of process capability based on regular updates:* Correction should only be undertaken if the manufacturing process average or standard deviation has shifted beyond the normal statistical significance, for either variable or attribute processes, and the population distribution is assumed to be normal. The distribution of the data should be symmetrical, with no skew. If not, the process should be investigated.

2.2.2 Determination of Standard Deviation for Process Capability

There are four different methods to determine the standard deviation σ of the population for process capability studies:

1. *Total overall variation:* All data are collected into one large group and treated as a single large sample where n is greater than 30.

2. *Within-group variation:* Data are collected into subgroups, and a dispersion statistic is calculated (the range). All ranges of each subgroup are averaged into an $\bar{R}$. The σ is calculated from an $\bar{R}$ estimator (d_2). This method is the basis for variable charts control limit calculations.

3. *Between-group variation:* Data are collected into subgroups, and an average ($\bar{X}$) is calculated for each subgroup. The standard deviation s of sample averages is calculated. The population σ is estimated from the central limit theorem equation $\sigma = s * \sqrt{n}$. This method can be used to obtain process capability from control chart limits.

4. *Moving-range method:* In this method, data are collected into one group of small numbers of data, over time. A range R is calculated from each two successive points. All ranges of each pair are averaged into an $\bar{R}$. The σ is calculated from an $\bar{R}$ estimator (d_2) for $n = 2$, which is equal to 1.128. Method 4 is

the preferred method for time series data and small data sets from low-volume manufacturing.

For processes that are in statistical control, these methods are equivalent over time. For processes not in control, only method 2 is insensitive to process variations of the average over time and is best recommended for testing new green materials. The σ estimate is inflated or deflated with method 1 and could be severally inflated/deflated with method 3. An example of a process out of control could be that one subgroup has a large sample average shift as opposed to smaller average shifts in the other subgroups. Another way to leverage the method 2 advantage to negate the effect of average shift is to use method 4, with the data spread over time.

2.2.3   Example of Methods of Calculating σ

Data for a production operation were collected in 30 samples, in three subgroups measured at different times. The four different calculations of σ are as follows.

Subgroup	Measurement	Subgroup Range R	Average	s
I	4, 3, 5, 5, 4, 8, 6, 4, 4, 7	5	5	1.56
II	2, 4, 5, 3, 7, 5, 4, 3, 2, 5	5	4	1.56
III	3, 6, 7, 6, 8, 4, 5, 4, 6, 6	5	5.5	1.51
Average of subgroups I–III		5	4.83	1.54
For the total groups		6	4.83	1.62
Moving range for each subgroup		Total	$\bar{R}$	σ
I	1, 2, 0, 1, 4, 2, 2, 0, 3	15	1.67	1.48
II	2, 1, 2, 4, 2, 1, 1, 1, 3	17	1.89	1.68
III	3, 1, 1, 2, 4, 1, 1, 2, 0	15	1.67	1.48
	Average moving range (MR)		1.74	1.54

Method 1. Total overall variation of 30 data points from three subgroups:

$$\sigma^2 = \frac{\sum_i (y_i - \bar{y})^2}{n-1} = \frac{\sum_i y_i^2 - \left(\sum_i y_i\right)^2 / n}{n-1} = [777 - (145)^2 / 30] / 29 = 2.626$$

$$\sigma = 1.62$$

Method 2. Within-group variation; $\bar{R} = 5$ ($n = 10$):

$$\sigma = \bar{R}/d_2(n = 10) = 5/3.078 = 1.62$$

Method 3. Between-group variation:

$$s_{\bar{x}} = s(5, 4, 5.5) = 0.764$$

$$\sigma = s * \sqrt{n} = 0.764 * \sqrt{10} = 2.415$$

Method 4. Moving-range method ($n = 2$). For each subgroup, obtain the average range between successive numbers:

Subgroup I	$\sigma = \bar{R}/d_2(n = 2) = 1.67/1.128 = 1.48$
Subgroup II	$\sigma = \bar{R}/d_2 = 1.89/1.128 = 1.68$
Subgroup III	$\sigma = \bar{R}/d_2 = 1.67/1.128 = 1.48$
For total groups (I–III)	$\sigma = \bar{\bar{R}}/d_2 = 1.74/1.128 = 1.54$

As can be seen above, the standard deviation σ of the overall 30 numbers was 1.62 (method 1). The 30 numbers comprised three subgroups (samples) with large shifts in sample averages. The closest indirectly calculated σ value was obtained by method 2, the within-group variations from the $\bar{R}$ estimator of σ, because it negated the average shifts. The moving-range method (method 4) was as much as 10 percent off, even when the full 30 numbers were used. The least accurate value was found by method 3, the between-group variation, which derived σ from distribution of sample averages and the conversion of sample to population σ. The number of subgroups (samples) was small and led to the largest error in σ determination.

2.2.4 Process Capability for Low-Volume Production

When the data required to determine process capability are not feasible to collect, because of cost or resource issues or production volume, reduced data can be used successfully to estimate process capability, provided that confidence is quantified in the data analysis. While 30 points of data are considered statistically significant, a smaller number of data values can be taken, using predetermined error levels and confidence goals, to obtain a good estimation of process average and variability.

The moving-range method provides for an alternate mechanism for estimated σ for a small amount of data, provided that data points are taken over time for both variable and attribute processes. Ten data points are required to provide an estimator for σ in the moving-range method.

2.3 Process Capability for Prototype New Green Parts

When prototype parts are acquired, such as those for green materials and processes, the testing methodologies should be made with alternate t and F tests, comparing the new green parts with the capability of the established nongreen parts that they are replacing, and the formulas being used are dependant on the sample sizes taken.

The formulas for combining s (small samples) or σ (large samples) from two distinct samples with varying sample sizes (n_1 and n_2) into one combined or pooled term are as follows:

For large samples (>30) of standard deviation σ_1, σ_2 and sample sizes n_1, n_2

$$\sigma_{\text{pooled}} = \sqrt{\frac{\sigma_1^2}{n_1} + \frac{\sigma_2^2}{n_2}} \tag{2.10}$$

For small samples (<20) of standard deviation s_1, s_2 and sample sizes n_1, n_2

$$s_{\text{pooled}} = \sqrt{\frac{s_1^2(n_1 - 1) + s_2^2(n_2 - 1)}{n_1 + n_2 - 2}} \tag{2.11}$$

To compare large samples to see if the difference between sample averages is significant, a test statistics z is generated:

$$z = (\overline{X_1} - \overline{X_2}) / \sigma_{\text{combined}} = (\overline{X_1} - \overline{X_2}) / \sqrt{\frac{\sigma_1^2}{n_1} + \frac{\sigma_2^2}{n_2}} \tag{2.12}$$

By repeating the above for the difference of small sample averages, t is calculated with ($n_1 + n_2 - 2$) degrees of freedom (DOF):

$$t = (\overline{X_1} - \overline{X_2}) / \left(S_{\text{pooled}} * \sqrt{\frac{1}{n_1} - \frac{1}{n_2}} \right) \tag{2.13}$$

$$t = (\overline{X_1} - \overline{X_2}) * \sqrt{\frac{n_1 n_2 (n_1 + n_2 - 2)}{n_1 + n_2}} / \sqrt{(n_1 - 1)s_1^2 + (n_2 - 1)s_2^2} \tag{2.14}$$

The use of the pooled standard deviation can then be expanded to the confidence limits based on the pooled degrees of freedom of $n_1 + n_2 - 2$. Note that the multiplication of S_{pooled} by the square root of the sample sizes is sometimes called the normalized S_{pooled}.

In the lead-free consortium phase I testing, a DoE was set up to determine whether defect-free soldering can be achieved with lead-free. The experiment is further explained in Sec. 2.4. With different

	Lead-Free			Tin-lead Baseline		
Sl.no.	PWBs	Visual Defect	Pull Strength	PWBs	Visual Defect	Pull Strength
1	5A	0	35.41	Pb 7	0	21.22
2	5B	0	31.83	Pb 8	0	22.27
3	6A	0	42.20	Pb 9	0	19.30
4	6B	0	30.77			
5	22A	0	27.96			
6	22B	0	31.49			
7	23A	0	29.01			
8	23B	0	34.98			
	Average		32.96			20.93

TABLE 2.3   Comparison of Pull Tests: Lead versus Lead-free for Consortium Phase I

materials and processes, 54 PCBs were soldered, and all solder joints were examined. Eight lead-free PCBs and three leaded PCBs were found to be defect-free by visual inspection. In addition, the leads of one IC on in each PCB were pulled until the solder joints broke. The distribution of the data is given in Table 2.3, showing that the average pull force for lead-free solder was 32.95 N, while the average pull force for the equivalent tin/lead solder was 20.93 N.

The data were analyzed using Microsoft Excel for t and F tests, since the sample size was small. To perform similar tests, use the Tools drop-down menu in Excel, and then select Data Analysis. If your system does not have Data Analysis installed, go to the Add-Ins button under Tools and install the Analysis ToolPak by checking the box and clicking OK. You will need to use the installation CD that came with Microsoft Office. Within Data Analysis, choose t or F tests. The data analysis is summarized in Table 2.4.

The first step is to perform the F test, to ascertain whether the standard deviations of the samples are significant to one another. If they are not, then the test to be used is the two-sample t test assuming equal variances, illustrated in Eqs. (2.13) and (2.14). If the F test shows unequal variances, different equations will apply that are not shown here. The results of the F test indicate an F value of 9.09 calculated below. This is less than the table value of 19.35 (indicated as the critical one-tail value in Excel). This value can also be obtained from Table 2.9 in this chapter for 95 percent confidence. When the calculated F value is less than the table value using the degrees of freedom specified (2 for tin/lead sample and 7 for lead-free samples), then the two samples are considered not significant to each other. Note that the calculated F test should always be greater than 1. The Excel analysis

	Samples	
	Lead-Free	Tin/Lead
95% Confidence interval for mean	32.96 ± 3.80	20.93 ± 3.74
Standard deviation	4.54	1.51
Variance	20.61	2.27
Pooled variance		16.54
Degrees of freedom	7	2
F test result (hypothesis: Sigma1 = Sigma 2)		9.09
F value from table (for $n_1 = 7$, $n_2 = 2$, and upper 5%)		19.35
F test probability $P(F <= f)$		0.10 or 10%
t test result (assuming equal variances)		4.37
t value from table (two-sided $\alpha = 0.025$, $r = 9$)		2.262
F test probability $P(T <= ft)$		0.00

TABLE **2.4**   Analysis of Pull Tests: Lead versus Lead-free for Consortium Phase I

will also return the probability of this situation of two samples occurring in nature or P ($F <=$ Table F). This probability is 10 percent and therefore considered to be not significant since it is larger than the 5 percent specified in the table.

As can be seen from Table 2.4, the t test assuming equal variances indicates that there is a significant difference in the average pull tests between lead-free and tin lead soldering PCBs that are free of visual defects. Table 2.1 gives the t_α value for two-sided distribution for $\alpha = 0.025$ and $r = 9$ as $t_\alpha = 2.262$. The resulting t_α from the experiment was 4.37 [using Excel or deriving from Eq. (2.16)], which is much higher than the table value of 2.262, indicating a significant difference in the average mean.

To show the manual calculations as used in this section, from Eq. (2.7), the confidence interval can be calculated for both samples:

$$95\% \text{ confidence of sample } 1 = t_{\alpha/2,\, n-1} * s/\sqrt{n}$$

$$= 2.365 \text{ (Table 2.1)} * 4.54/\sqrt{8} = 3.80$$

$$95\% \text{ confidence of sample } 2 = t_{\alpha/2,\, n-1} * s/\sqrt{n}$$

$$= 4.303 \text{ (Table 2.1)} * 1.51/\sqrt{3} = 3.74$$

$$F \text{ test} = \text{variance sample } 1/\text{variance sample } 2 = 9.09$$

From Eqs. (2.14) and (2.16), the t value can be calculated:

$$S_{pooled} = \sqrt{(\{[\text{Variance}_1 * (n_1 - 1)]}$$
$$+ ([\text{Variance}_2 * (n_2 - 1)]\}/(n_1 + n_2 - 2))$$
$$S_{pooled} = \sqrt{\{(20.61 * 7 + 2.27 * 2)/9\}} = 4.066$$

Pooled variance $= S^2_{pooled} = 16.54$

Normalized $S_{pooled} = S_{pooled} * \sqrt{1/n_1 + 1/n_2}$

$$= 4.066 * \sqrt{1/8 + 1/3} = 2.753$$

Calculated t for 2 samples = difference in sample average/ normalized S_{pooled}

Calculated t for 2 samples $= (32.95 - 20.93)/2.753 = 4.37$

2.4   Statistical Methods Conclusions

It was shown in this chapter how to handle the common problem of applying quality methodology to small as well as large production volumes. Statistical tools such as the moving range and the Z, t, F, and χ^2 distributions can be used to quantify the attributes of the population distribution for average and standard deviations based on samples taken. Many examples were given to demonstrate sampling techniques and their relationship to populations. Process capability was demonstrated with formulas, examples, and case studies. Finally the process capability applications in short- versus long-term green materials use were also shown, with examples and strategies for handling process capability in the prototype as well as long-term production.

2.5   Design of Experiments

The concepts of the design of experiments, alternately known as *robust design* or *variability reduction*, have been used to reduce some of the sources of manufacturing variation or manipulate a design toward its intended performance. In addition, DoE is an excellent methodology to investigate green materials and process alternatives.

In Sec. 2.5.1 the definition of DoE is given, as well as the expectations of proactive improvement of the product and process design. The reasons for DoE are discussed, including the effects of noise and other external and internal conditions that contribute to the variability of products and processes.

In Sec. 2.5.2 these techniques are introduced with an algorithm for conducting a DoE project and selection of the green quality characteristic.

In Sec. 2.5.3 these tools are presented with case studies for each. They include graphical and statistical analysis of the average and the variance of the quality characteristic.

In Sec 2.5.4 the use of DoE tools and techniques are illustrated for various phases of green material and process selection.

2.5.1 DoE Definitions and Expectations

Design of experiments is a systematic method for determining the effect of factors and their possible interactions on a design or a process toward achieving a particular output of the quality characteristic(s). It is used to quantify the source and resolution of variation and the magnitude of the error when comparing the average of the quality characteristic to the target.

By using DoE techniques, a design or a process can be manipulated to provide a target or minimum/maximum performance of the quality characteristic average or reduce its variability or do both. This is accomplished by setting factors that affect the quality characteristic to predetermined levels and analyzing the output sets of factorial, partial factorial, or orthogonal experimental arrays.

The objectives of DoE are to adjust the quality characteristics (or design or process output) to the optimum performance by properly choosing the best combination of factors and levels. This is accomplished by collecting maximum information from the DoE results while using minimum resources. The factors can be categorized to determine which factors affect the average, affect the variability, affect both the average and the variability, or have no effect on quality characteristics. The results of a DoE can be one of the following:

1. Identify the most important factors that influence the quality characteristic.

2. Determine factor levels for the important factors that optimize desired quality characteristics (output responses).

3. Determine the best or most economic setting for factors that are not important.

4. Validate (confirm) responses and implement in production or design.

The success of an experiment is not determined solely by just achieving the desired quality level. Important information about the design or the manufacturing process can be gleaned from any experiment. This information can be put to use in future experiments or through using more traditional quality improvement processes such as total quality management (TQM). Information gained from DoE can be listed as follows:

- The factors that are significant for influencing the quality characteristic average, reducing variability, or doing both and which of those factors are not significant. If none of the factors are found to be significant, then the design of the experiment has to be repeated to include factors or levels not previously considered.

- The proper balance between average shift from target and variability reduction by choosing the proper factor levels. The choices of certain factor levels can shift the average, while other choices can reduce the variability, or can do both. While good results can be obtained by moving the average to the maximum or minimum possible or to achieve a target for the design, this action can be tempered by selecting alternate factors and levels to achieve the greatest robustness in reducing variability.

- The predicted experiment outcome can be determined when the design or factors are set to the specified levels. Confidence intervals and the expected error can also be shown for the predicted outcome.

- The goodness of the DoE design and the proper selection of factors and levels can be evaluated by statistical analysis.

2.5.2   DoE Techniques

DoE is best characterized as making several assumptions about the design or the process being studied, quantifying these assumptions by the choice of factors and levels, and then running experiments to determine if these assumptions are valid. It is a mix of several tools that have been developed to optimize performance, based on statistical analysis, significance tests, and error calculations.

DoE uses statistical experimental methods to develop the best factor and level settings to optimize a green process or a design. Some of the statistical methods have been simplified by the use of specialized software analysis packages.

Steps in Conducting a Successful DoE

Conducting a DoE experiment involves using many of the tools common to TQM. It is always advantageous to form a team to perform the tasks of designing the experiments and interpreting the results. Teams can have shared experiences in green design and can achieve a broad consensus on different approaches to the DoE and the problem being analyzed.

The success of a green DoE project is dependent on selecting the proper team members, identifying the correct factors and levels, focusing on optimizing and measuring the quality characteristics, and analyzing the results. Steps in performing a successful robust design of experiments are as follows:

- *Problem definition*: The first task in performing a DoE project is to outline the goals of the project and to define the quality characteristic(s) of the process or the design to be optimized. Although only one characteristic can be optimized at a time, many characteristics can be measured while performing the experiments, and analyzed separately, from the same experiment matrix. The final level selection can be a mix of the recommended factor and level settings, depending on the compromise of the different objectives of each quality characteristic.

- *Design space:* Creating the boundary of the product or design to be optimized is important. The experiment should not be constrained to a small part of the design, hence not having the opportunity to study the interactions of the different parts of the total design. On the other hand, the experiment should not be all-encompassing in attempting to optimize a wide span of product design steps or processes. Ideally, the total design should be analyzed, and a compromise made in developing a plan for a succession of DoEs, with each additional experiment providing more information about the design to be optimized.

- *Team creation and dynamics:* A project team should be selected to conduct the experiments and perform the analysis. The team should be composed of those knowledgeable in the materials, product, and process, and should solicit inputs from all parties involved in the design to be optimized. It is not necessary to have in-depth technical understanding of the science or technology of the problem, but the team members should have experience in similar or previous designs. Knowledge in statistical methods, and in particular DoE techniques, should be available within the team, through either a statistician or someone having received training or experience in DoE.

- *Factor and level selections:* DoE can be performed in two approaches: One method is to select a large number of factors and to use a screening experiment, usually a saturated design (to be explained further), to narrow down the factor selections. Then a follow-on experiment, preferably a full factorial experiment, is used to complete the selection of the optimal factors and their levels. The second method is to have the team members consider this DoE project as a single opportunity to try out as many as possible factors, levels, and combinations of both because of the lack of time or resources available. In this case, partial factorial experiments are used, with some assumptions as to the relationships of factors, to maximize the benefits, resources, and time spent on a single experiment.

- *Brainstorming techniques:* These should be used to select the number of factors and the different levels for each factor. The selection process should outline factors that are as independent as possible from other factors, and hence are additive in controlling the quality characteristic(s) to be optimized. This is important in reducing the interactions of factors, which are difficult to quantify statistically.

  An example of selecting independent factors and reducing their interactions is the case of an infrared conveyorized oven for the reflowing of surface mount technology (SMT) PCBs. The reflow process is characterized by three factors: the ramp-up of temperature to the solder melting stage, the maximum temperature level reached during reflow, and the time during which the temperature remains above the solder liquid state, usually called time above liquidus (TAL). There are several heater zones in the reflow oven, and the oven temperature can be controlled by setting the zones on the top and bottom of the reflow oven to predetermined levels, as well as varying the conveyor speed. Choosing temperature zones and the conveyor speed as the factors for a reflow experiment would result in strong interaction between the factors. The proper choice of factors would be the ramp-up temperature rate, the TAL, and the maximum reflow temperature. The factor levels selected should be achieved by actually experimenting with the temperature zones and conveyer speed to reach the desired levels in the experiment.

- *Level selection:* Good selection of the levels for each factor used in the experiments is important in achieving the proper design space. Levels that are either too close together or too far apart in value should not be selected, because they do not represent a continuum of the impact of the factor on the measured characteristic. Level selection should follow these guidelines:
  1. Three level designs could be chosen if the project team is confident that the current design is performing adequately, but needs to be improved. The current level should be in the center of a 20 percent span represented by the other two levels. In this manner, the DoE can help in finding a more optimized operating set of factor levels in the design space.
  2. Two levels could be selected if there is little confidence in the adequacy of the current design, based on the collective

judgment of the team. By choosing two levels, more factors can be tested within a small number of experiments, as will be demonstrated later. In addition, the direction of better design performance can be ascertained for future DoE.

3. Multiple level factors should be chosen for survey experiments. In these DoEs, a team can select many new technologies or materials within one factor to identify which one can perform best in the design. The number of multiple levels should be close to squares of 2 or 3 levels, such as 4, 8, or 9 levels. They are easier to perform since they fit easily into the set of predetermined experiment arrays.

4. The selected levels should be well within the operating range of a working characteristic within the design space. In the soldering reflow experiment mentioned earlier, the combination of temperature factors and levels should not result in having components soldered beyond their maximum temperature and time exposure specifications.

- *Experiment arrays:* Most DoE experiments use a set of standard orthogonal arrays available to conduct the experiment, with two or three levels. There are only certain combinations of factors and their levels available to perform the experiment. Compromise might be necessary to achieve economy in DoE by selecting a given number of factors and levels that can fit within one of the orthogonal arrays. There are only a small number of these arrays of two and three levels, and their size increases geometrically with the number of factors selected.

- *Orthogonal arrays:* Conducting the experiments is based on the selected orthogonal arrays. The arrays are arranged in terms of the number of experiments, factors, and levels. The experiments should be conducted in a random order from the array matrix. The measurements of the characteristic to be optimized could be repeated using various scenarios, depending on the variability considerations of the design.

- *Data analysis:* Once the experiments are performed, the data can be analyzed graphically to determine the optimal settings of levels of significant factors. In addition, statistical analysis can be performed to determine the significance of each factor's effect on the quality characteristic, through the use of analysis of variance (ANOVA). Important factors can be set to the proper level, and least-significant factors can be ignored, or set to the most economic conditions.

  - *Graphical analysis of the data.* This is sufficient to determine the best factor setting to adjust the design average to target and reduce design variability. *Statistical analysis* of data provides more details on the probability of the effect of each factor on the characteristic measurement. In addition, statistical analysis can quantify the usefulness of the DoE project: low significance of the total experiment usually results from the lack of significant factors. In this case, the experiment is not providing useful guidance to the design team, and it should be repeated with additional or different factors and levels.

- *Prediction and factor and level selection:* Once the graphical and statistical analysis is completed, the characteristic value can be predicted based on the choice of factor levels. These choices could be a compromise between setting the design characteristic average to the target value and reducing variability. A recommended factor level might cause variability to be reduced, yet at the same time the process average will be shifted from target. Another case occurs when multiple characteristics are to be optimized using one experiment with many separate output measurements and data analysis. For example, a DoE could be performed to design a new plastic material to be injection-molded. The material and process design can have several desired characteristics including modulus of elasticity, density, amount of flash after injection, gel time, flow rate, and free rise density. A DoE could be designed using an orthogonal array that determines what ratios and compositions of raw materials are to be used, as well as the injection-molding machine parameters. Measurement of all the desired characteristics will be performed, and then the data are analyzed to determine the best set of raw material ratios for each characteristic. A compromise of all recommended factor levels will have to be made to achieve the best overall plastic product.

- *Confirmation experiment:* Once all the choices and predictions of the DoE have been agreed upon, a confirmation experiment run should be made before final adoption of the design decision, to verify the analysis outcome. This confirmation will test out the entire robust design process before full implementation takes place. In manufacturing, the newly adjusted process should continue to be monitored through statistical quality control methods for a 6-month minimum time period, before any attempts be made to further increase the robustness of the process by launching another DoE.

Types of DoEs Using Orthogonal Arrays

The arrays most commonly used in the design of experiments are the orthogonal arrays (OAs). These arrays are balanced: there are an equal number of levels for each factor in the experiments. The behavior of each factor level can be studied while other factors are changing their levels. This technique results in an array matrix with n columns and $n + 1$ experiments with two level n factors.

Orthogonal arrays allow for measuring the effect of varying many factors at the same time by measuring factor interactions, through several techniques:

1. Full factorial DoEs are used to evaluate the effects of all factors and their interactions. For every number of factor columns n, there are at least $n - 1$ interactions columns to be considered, and for m levels, there are m^n experiments. For example, if four factors are considered (A, B, C, and D) with two levels, there are 11 interactions, such as AB, AC, AD, BC, BD, CD, ABC, ABD, BCD, and $ABCD$, and $16 = 2^4$ experiments. The levels in the interaction columns are derived from the multiplication of the levels of the originating factors, using an exclusive-OR (XOR) logical formula. For full factorial designs, the number of experiments increases geometrically as the number of factors increases. Fractional factorial DoEs provide a cost-effective way of determining the significance of selected factor interactions. A fractional factorial DoE uses a portion of the full factorial columns to estimate main factor effects and their interactions. The unused interaction columns are then assigned to other factors, resulting in the condition called *confounding*, where the assigned factor could be confounded with the interaction that is normally found in the column. By selectively choosing where to confound, a fractional factorial DoE could be used to study more factors, but with fewer experiments than full factorial.

2. Saturated design DoEs allow all the columns in the OA to be assigned to different factors. They represent a minimum set of experiments for the number of factors considered. They are called *screening designs*, because they are commonly used to whittle down the number of factors quickly through a smaller DoE; then a full factorial DoE can be performed on the remaining factors. The assumption in saturated designs is that interactions are small and can be ignored compared to the main factor effects.

Two-Level Orthogonal Arrays

The most commonly used two-level orthogonal array is the L8. It is an eight-experiment array, sometimes referred to as 2^7, which has seven columns to be used as factors (1 through 7 or A through G), and each factor is to be considered at two levels (1 and 2), as shown in Table 2.5. The symbols for the factors are given for the top two rows as saturated design, for the next row for full factorial, and for the

Experiment Number	Factor Symbols						
	1	**2**	**3**	**4**	**5**	**6**	**7**
	A	*B*	*C*	*D*	*E*	*F*	*G*
	A	*B*	*AB*	*D*	*AD*	*BD*	*ABD*
	A	*B*	*AB*	*D*	*AD*	*BD*	*G*
1	1	1	1	1	1	1	1
2	1	1	1	2	2	2	2
3	1	2	2	1	1	2	2
4	1	2	2	2	2	1	1
5	2	1	2	1	2	1	2
6	2	1	2	2	1	2	1
7	2	2	1	1	2	2	1
8	2	2	1	2	1	1	2

TABLE 2.5 L8 Orthogonal Array

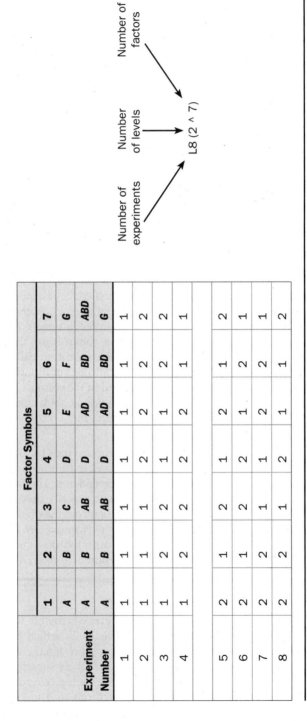

Number of experiments

Number of levels

Number of factors

L8 (2 ^ 7)

bottom row for partial factorial design using resolution IV. Note that there are three uses for the L8:

1. Use as full factorial array to check three factors (A,B,D) at two levels and four interactions [$C(AB)$; $E(AD)$, $F(BD)$, $G(ABD)$].

2. Use as a saturated (screening) design for up to seven factors at two levels and no interactions. When this array is used for saturated designs, factors should be assigned according to potential significance as follows: the most important factors should be the assigned to the primary columns A, B, D; column G, which confounds with the three-way interaction ABD, should be assigned next. For the last three factors to be assigned, use the columns C, E, F which confound with two-way interactions of the primary factors.

3. Use the L8 as a partial factorial design with three primary factors assigned to columns A, B, and D; a fourth factor G; and interactions C, E, F. There are several confounding and missing interactions in this application of L8: factor G confounds with the three-way interaction ABD, and two-way interactions of the three primary factors A, B, D with factor G are missing and assumed to be insignificant. Factor G is assumed to be independent of the previous three factors.

All eight experiment lines in an L8 can be repeated as necessary to establish average and variability analysis of the quality characteristic(s), and to obtain a statistically relevant sample. A simple rule is to use 30 values minimum for assuming a population distribution. In this case the L8 should be repeated 4 times. For large processes with many different factors such as an IC manufacturing line, the process can be divided into segments and each segment can be optimized individually with a DoE. It is much easier to conduct two L8 experiments than a single large experiment such as L32.

The use of the L8 as a saturated design with seven independent factors contrasts with their full factorial use. A full factorial design with seven factors would require a DoE with an L128 (2^7) experiment. What is gained by many fewer experiments in the saturated design (L8) is offset by its inability to calculate interactions as in the full factorial designs (L128).

The balance of the orthogonal arrays can be shown with the L8. For each particular level in a column, all the other levels in the other columns are rotated through their values. Experiments 1 through 4 have column A with level 1 only, while the levels in columns B through G contain both levels 1 and 2, in equal numbers of 2 each. The balance of the orthogonal arrays allows for a simple solution for the values of the factors in the experiments using Cramer's rule. In each array, there are n unknown variables that can be solved in $n - 1$ equations. The L8 can be represented with seven unknowns and eight simultaneous equations, and thus each variable (factor) can be solved for. The solution to the eight equations, sometimes called the *expected value* (EV), includes the equation constant (the experiment average) and the contribution of each factor depending on the level selected in the following form:

EV = average of all experiments + [1 or −1] * contribution of factor A + [1 or −1] * contribution of factor B + · · · + [1 or −1] * contribution of factor $A * B * C$

The next higher two-level array is an L16. The L16 includes 16 experiments and 15 columns at two levels. The L16 can be thought of as two L8's stacked on top of each other, with additional columns used for a 4th primary factor and its interactions. It is not necessary to choose all available columns to be included in the experiment: An L16 experiment can be performed with 10 factors in saturated design, and the other factors (array columns) can be left unassigned. This does not jeopardize the utility of the experiment, since the analysis of the effect of the 10 factors on the output characteristic is valid. The remaining factors could be used for calculating some of the interactions, according to the assignment of the main factors.

Three-Level Orthogonal Arrays

Three-level orthogonal arrays are popular in process improvements. Most current operations can be improved with DoE using the current process value as the middle level, and then extending 20 percent for above and below the current value for the other two levels. Three-level graphical analysis of the relationship between the factors and the quality characteristic(s) can be plotted using three points, and hence any curvature of the data can be shown, versus the straight line of the two-level experiments. Three-level columns have two two-way interactions each, so that factors A and B have two interactions AB and BA.

The smallest three-level orthogonal array that can be used is the L9, shown in Table 2.6. The top two rows are two different factor symbols commonly used for saturated design, and the bottom row is the factor symbol for full factorial design. The three-level orthogonal arrays, such as L9, have two uses similar to the L8:

1. Use as full factorial design to check two factors (A, B) at three levels and all their interactions [one two-way interaction with two columns $C(AB)$ and $D(BA)$].

2. Use as saturated (screening) design to check up to four factors (A, B, C, D or 1, 2, 3, 4) at three levels.

A partial factorial design is not possible in an L9. All nine experiment lines are repeated as necessary to establish average and

	Factor Symbols				
	A	*B*	*C*	*D*	
Experiment	1	2	3	4	
Number	*A*	*B*	*AB*	*BA*	Results
1	1	1	1	1	Y1
2	1	2	2	2	Y2
3	1	3	3	3	Y3
4	2	1	2	3	Y4
5	2	2	3	1	Y5
6	2	3	1	2	Y6
7	3	1	3	2	Y7
8	3	2	1	3	Y8
9	3	3	2	1	Y9

TABLE **2.6**   L9 Orthogonal Array

variability analysis of the quality characteristic(s). The use of the L9 as a saturated design of 4 factors and 9 experiments can be contrasted with the full factorial design of 4 factors and 81 (3^4) experiments.

Interaction and Linear Graphs

Interaction occurs when one factor modifies the conditions of another. If this is deemed significant, the interaction should be derived from its own column in the array, and no factor should be assigned to this column.

There are four interaction columns in orthogonal array L8. The interaction of columns 1 (factor A) and 2 (factor B) can be found in column 3 (C = A*B), forming an exclusive-OR relationship in the levels for column 3. These relationships can be grouped into primary factors and interaction factors. For example, in the partial factorial design of L8, columns 1(A), 2(B), 4(D), and 7(G) are primary factors in array L8, while the remaining columns are due to the interactions of the first three primary factors.

Interactions have caused much confusion for DoE teams. If an interaction is to be considered, fewer primary factors can be used which reduces the utility and economy of orthogonal arrays. For example, L27, which is 13 columns at three levels, can be used as a full factorial array with three primary factor columns 1, 2, and 5; six two-way interactions 3, 4, 6, 7, 8, and 9; and four three-way interactions

10, 11, 12, and 13. The number of primary factors in the L27 can increase to seven factors (1, 2, 5, 10, 11, 12, and 13). That would leave three two-way interactions in six columns, two columns for each interaction (3, 4 then 6, 7 and 8, 9). If more than seven factors are to be used when considering an L27, then they should each be assigned to a column where the one of the three interactions is considered insignificant. An example of an L27 array for lead-free is shown in Table 2.12.

Interaction represents a mathematical value of the effect of one factor on others. If an assigned factor confounds an interaction column, then the analysis of the effect of that factor could be either negated or amplified by the interaction effect. In actuality, the effect of interactions is usually much smaller than expected. For DoE teams concerned with the confusion of interactions, noninteracting orthogonal arrays such as L12 (otherwise known as the Plackett and Burman design), L18, or L36 could be used. In these arrays, any third column does not confound the interaction of any two columns. L12 is a two-level array, while L18 is a combination two-level and three-level factors.

2.5.3   DoE Analysis Tool Set

The DoE analysis tool set consists of using graphical as well as statistical analysis to determine which individual factors are significant, and how to set the quality characteristic to its design goal or reduce its variability. The graphical analysis takes advantage of Cramer's rule of the solution of simultaneous equations to solve for each value of factor level. In the L9 OA in Table 2.6, it takes nine experiments to perform a solution of four factors at three-level saturated design. The average of the results of the first three experiments $Y1$, $Y2$, and $Y3$ is the average performance of the product or process due to selecting level 1 of factor A, while the other factors negate themselves by averaging out their levels. The average of $Y2$, $Y5$, and $Y8$ is the effect of selecting level 2 of factor B. In this manner, the average of all 12 possible combinations (factors A, B, C, and D and their levels 1, 2, and 3) is examined in terms of attaining the best result for the product or process specifications. For an L9 with n repetitions, the level values for factor A can be calculated as follows:

$$A1 = \frac{\sum \frac{Y1 + Y2 + Y3}{n}}{n * 3} \qquad A2 = \frac{\sum \frac{Y3 + Y4 + Y5}{n}}{n * 3} \qquad A3 = \frac{\sum \frac{Y6 + Y7 + Y8}{n}}{n * 3} \qquad (2.15)$$

The data can be plotted graphically so the intended results of maximum, minimum, or targeted quality characteristic values can be used to manipulate the design to the intended or "expected" values.

The expected value (EV) of the DoE output is the result of applying all the recommended levels based on maximum, minimum, or target values. This is constructed from the overall experiment average, and then the contribution of each recommended level is added to the EV. The contribution is the recommended level value minus the experiment average. The contribution of interactions can be calculated from the selected levels of primary factors.

The EV value is usually calculated for significant factors only. The significant factors are determined by performing the F test using the ANOVA analysis in the next section. The contribution of nonsignificant factors could be lost within the error of the experiment (the confidence interval of EV). If the selected factor levels are within the experiment design as one of the experiment lines, the expected value should equal the value attained by the experiment line, and no calculations are necessary. All expected values are bounded by the confidence interval of the error, as mentioned in earlier sections of this chapter.

Analysis of DoE Data with Interactions: Lead-Free Solder Material and Process Selection Using L8

In this example, a DoE was used to determine the best selection of factors and levels for lead-free soldering. The experiment was designed with three factors at two levels: surface finish, paste type, and whether to use nitrogen or air in the reflow atmosphere. It was determined to conduct this experiment as a full factorial to understand the effect of factors and their interactions.

The L8 experiment was performed with three primary factors being considered, in a full factorial design, as shown in Table 2.7. Of the seven columns, three were assigned to the primary factors: A (finish), B (paste), and D (atmosphere). The remaining four columns were used to measure the interaction of the selected factors, using the interaction scheme shown in Table 2.5, with three two-way interaction columns: C (columns $A * B$ or finish $*$ paste), E (columns $A * D$ or finish $*$ atmosphere) and F (columns $B * D$ or paste $*$ atmosphere). The fourth column was a three-way interaction column G (columns $A * B * D$ or finish $*$ paste $*$ atmosphere). The primary factors consisted of using two different finishes (OSP or ENIG), two solder paste types (A or B to protect the identity of the material supplier companies) and nitrogen (N) or air in the reflow atmosphere.

Two sets of eight PCBs were made for each experiment line for a total of 16 sets. The PCBs were examined visually, and defects were identified and recorded according to industry standards inspection manuals. The 0.1 defect was actually a zero defect but was recorded as 0.1 to easily calculate the variability analysis, if desired.

The graphical analysis of data is given in Figure 2.1, using software developed by a former graduate student of the author, Val Carvalho. Two of three primary factors, paste and atmosphere, have a greater effect on the design than the third factor (surface finish), as will be shown with statistical analysis in later sections. In addition, only one interaction was larger than the rest, which was the interaction of surface finish $*$ paste supplier. This shows that depending on the

	Factor	Level 1	Level 2
A	Finish	OSP	ENIG
B	Paste	A	B
C	F x P	1	2
D	Atmosphere	Air	Nitrogen
E	F x A	1	2
F	P x A	1	2
G	F*P*A	1	2

A	B	C	D	E	F	G	Replication 1	Replication 2
Finish	Paste	F x P	Atmosphere	F x A	P x A	F*P*A	Visual Defects	Visual Defects
OSP	A	1	Air	1	1	1	0	11
OSP	A	1	Nitrogen	2	2	2	0	0.1
OSP	B	2	Air	1	2	2	82	58
OSP	B	2	Nitrogen	2	1	1	30	0
ENIG	A	2	Air	2	2	2	29	1
ENIG	A	2	Nitrogen	1	2	1	3	5
ENIG	B	1	Air	2	2	1	44	6
ENIG	B	1	Nitrogen	1	1	2	0	2

TABLE 2.7 Lead-free Solder Experiment Factor Selection

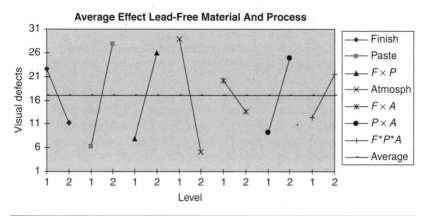

FIGURE 2.1 Lead-free material and process DoE graphical analysis.

paste material supplier selection, the PCB surface finish type might be important. This strong interaction indicates that the surface finish and the paste material supplier should be treated as one combination. The statistical analysis to be explored in the next section could determine which of these factors or interactions are significant.

Table 2.8 shows an analysis of the effects of the different factor levels on the visual defect rate. The average of all experiments is 16.94, or approximately 17 defects per PCB. Factor A (finish) level calculations are given below:

Factor A (finish) level 1 (OSP) = $(0 + 0 + 82 + 30 + 11 + 0.1 + 58 + 0)/8$
$= 181.1/8 = 22.6375$ defects

Factor A (finish) level 2 (ENIG) = $(29 + 3 + 44 + 0 + 1 + 5 + 6 + 2)/8$
$= 90/8 = 11.25$ defects/PCB

Factor	Level 1	Level 2	Average	Selected Level
Finish	22.6375	11.2500	16.9438	
Paste	6.1375	27.7500	16.9438	1
$F \times P$	7.8875	26.0000	16.9438	
Atmosphere	28.8750	5.0125	16.9438	2
$F \times A$	20.1250	13.7625	16.9438	
$P \times A$	9.1250	24.7625	16.9438	
F^*P^*A	12.3750	21.5125	16.9438	
Expected value		–5.7938		

TABLE 2.8 Lead-free Solder Experiment Level Selection

Obviously, if a factor is proved to be significant in ANOVA statistical analysis in the next section, then OSP is the better level selection for lower defects. If surface finish proves to be not significant using ANOVA, then there is no difference in defects when using OSP or ENIG for surface finish.

If paste level 1 (material supplier A) is selected and nitrogen is used in the reflow process, this will result in a very low defect PCB, shown with an expected value of –5.79 or –6 defects per PCB. This negative rate indicates that the confidence interval of the visual defects is very high since the resulting term cannot be negative. This is verified in the experimental results in Table 2.7, where there are two instances of paste *A* and nitrogen in the experiment lines 2 and 6. In these instances the visual defect rate average for the two replications PCBs were 0 and 4.

Statistical Analysis of DoE

Statistical analysis of DoE is based on the analysis of variance (ANOVA), which is a method of determining the significance of each factor in terms of its effects on the output quality characteristic(s). The ANOVA apportions the total effect of the output characteristic average and variability to each factor in the orthogonal array. The significance test is based on the F distribution, which is a ratio of the degrees of freedom for the factor divided by the degrees of freedom for the error. The least significant factors are pooled (lumped together) as the error of the experiment, since they are not important in affecting the output characteristic.

The terms for determining the ANOVA table for n total values in the DoE are given as follows:

$$\text{Total sum of the squares } (SS_T) = \Sigma(Y_i - Y_{\text{average}})^2$$
$$= \Sigma Y_i^2 - (\Sigma Y_i)^2/n \qquad (2.16)$$

Sometimes $(\Sigma Y_i)^2/n$ is called the correction factor.

$$\text{Total sum of squares for each factor } (SSF) = (\Sigma Y_{\text{level 1}})^2/n_{\text{level 1}}$$
$$+ (\Sigma Y_{\text{level 2}})^2/n_{\text{level 2}} + \cdots - (\Sigma Y_i)^2/n \qquad \text{(repeat for as many levels)} \qquad (2.17)$$

Degrees of freedom (DOF) $\qquad (2.18)$

DOF total	= no. of data points − 1
DOF orthogonal array	= no. of experiments − 1
DOF factor	= no. of levels − 1
DOF error	= total DOF = DOF of significant factors and interactions
DOF interaction	= product of DoF of each factor
	= 1 for two-level factor interactions
	= 2 for three-level factor interactions

Variance $V = SS/\text{DOF}$ (also called mean square deviation or MSD)

$$V_T = \sigma^2_{\text{total experiment}} \tag{2.19}$$

$$F \text{ ratio for each factor} = V_F / V_{\text{Error}} \tag{2.20}$$

$$\text{Modified sum of squares for each factor } SS'_F = SS_F - V_{\text{Error}} * \text{DOF}_F \tag{2.21}$$

$$\text{Percentage contribution } (p\%) = SS'_F * 100/SS_T \text{ Percent} \tag{2.22}$$

The F test values are given in Table 2.9, for a confidence level of 95 percent and the DOF of the factor versus the DOF of the error. The F test is used to determine the significance of the calculated variances. It is a ratio of the factor variance over the error variance. The error of the DoE could be obtained from any of the following:

1. Replicate the whole experiment, generating error due to repetition.
2. For single value DoE results data, the smallest factors or interactions (with the smallest SS_F) can be used as the error, especially higher-order interactions.
3. Replicate the center point of the design space of the experiment.
4. Replicate some points of the experiments, such as the endpoints of the design space.

For a given confidence level, The F test determines whether the effect of a factor is due to chance or due to the factor itself (the factor is deemed significant). If a factor's F ratio value is less than the value in the F table given the DOF of factor and error, then it is deemed not significant and can be pooled into the error. The F ratios are then recalculated, and the F test is redone on the remaining factors. When a factor is significant to less than 0.05 (or the confidence is greater than 95 percent), then the probability of this factor affecting the experiment happens 5 percent by chance or once every 20 times. Since this is remote in nature, the factor must be significant, and hence it affects the experiment outcome.

The last two terms (2.21 and 2.22) in the ANOVA table were developed by Genishi Taguchi, a noted DoE advocate, to simplify the pooling process. Instead of using the significance based on the F table as the source of pooling, Taguchi suggested to pool a factor if its percent contribution is less than 5 percent.

Statistical Analysis of Lead-Free Experiment

For the lead-free experiment, the initial ANOVA table is constructed in Table 2.10. An example is given later of how to calculate the sum of the squares for factor A.

DOF Error	DOF Factors								
	1	2	3	4	5	6	7	8	9
1	161.4	199.5	215.7	224.6	230.2	234.0	236.8	238.9	240.5
2	18.51	19.00	19.16	19.25	19.30	19.33	19.35	19.37	19.38
3	10.13	9.55	9.28	9.12	9.01	8.94	8.89	8.85	8.81
4	7.71	6.94	6.59	6.39	6.26	6.16	6.09	6.04	6.00
5	6.61	5.79	5.41	5.19	5.05	4.95	4.88	4.82	4.77
6	5.99	5.14	4.76	4.53	4.39	4.28	4.21	4.15	4.10
7	5.59	4.74	4.35	4.12	3.97	3.87	3.79	3.73	3.68
8	5.32	4.46	4.07	3.84	3.69	3.58	3.50	3.44	3.39
9	5.12	4.26	3.86	3.63	3.48	3.37	3.29	3.23	3.18
10	4.96	4.10	3.71	3.48	3.33	3.22	3.14	3.07	3.02
15	4.54	3.68	3.29	3.06	2.90	2.79	2.71	2.64	2.59

20	4.35	3.49	3.10	2.87	2.71	2.60	2.51	2.45	2.39
25	4.24	3.39	2.99	2.76	2.60	2.49	2.40	2.34	2.28
30	4.17	3.32	2.92	2.69	2.53	2.42	2.33	2.27	2.21
60	4.00	3.15	2.76	2.53	2.37	2.25	2.17	2.10	2.04
∞	3.84	3.00	2.60	2.37	2.21	2.10	2.01	1.94	1.88

TABLE 2.9 F Table Value for 95 Percent Confidence

Factor	Pooled	DOF	SS	MS	F	SS'	% Contribution
Finish	n	1	518.701	518.701	2.165	279.137	2.98
Paste	n	1	1868.401	1868.401	7.799	1628.838	17.39
$F \times P$	n	1	1312.251	1312.251	5.478	1072.688	11.45
Atmosphere	n	1	2277.676	2277.676	9.508	2038.113	21.76
$F \times A$	n	1	161.926	161.926	0.676	-77.638	-0.83
$P \times A$	n	1	978.126	978.126	4.083	738.562	7.88
$F*P*A$	n	1	333.976	333.976	1.394	94.412	1.01
Replication error		8	1916.505	239.563			
Pooled error		0	0.000	—			
Total error		8	1916.505	239.563		3593.447	38.36
Total		15	9367.56	624.504		9367.559	100.00

TABLE 2.10 Lead-free Solder Experiment ANOVA

$$SS_A = (\Sigma Y_{\text{level 1}})^2 / n_{\text{level 1}} + (\Sigma Y_{\text{level 2}})^2 / n_{\text{level 2}} - (\Sigma Y)^2 / N$$

$$SS_A = 1/8[(0 + 0 + 82 + 30 + 11 + 0.1 + 58 + 0)^2 \\ + (29 + 3 + 44 + 0 + 1 + 5 + 6 + 2)^2] - 271.1^2/16$$

$$SS_A = 1/8[(4099.651 - 1012.5) - 4593.45 = 518.701$$

To calculate the F ratio, each variance must be compared with the error variance. The error variance is based on the repetition of the experiment with a degree of freedom = 8. It is calculated from subtracting the sum of the squares of all factors from the total sum of the squares, and is equal to 1916.505. A factor is significant when its F value is greater than the F value from the table for DOF(factor) = 1 and DoF(error) = 8, which is equal 5.32. If the F value of a factor is less than the F value from the table for 95 percent confidence, then this factor is deemed not significant and pooled with the error. It is important to pool one factor at a time, starting with the factor with the lowest F value, since the F value of the remaining factors changes with each pooling step. The process is continued until no greater significance is achieved and only significant factors remain in the ANOVA table.

In this manner, the completed pooling steps are shown in Table 2.11, with only two factors that are significant, the solder paste material supplier and reflow atmosphere. These two factors should exceed the table value for 95 percent confidence and 1 degree of freedom for the factor and 13 degrees of freedom for the error which is 4.67. (Factor finish is slightly below at 4.65.) The remaining primary factor, surface finish, is not significant; neither are any of the interactions. The factors can be ranked in importance according to the percent contribution: atmosphere (nitrogen) is most important, followed by paste supplier A. The total percent contribution of the error is 64 percent, indicating an abundance of variability in the data. If the error percent is greater than 50 percent, the significance of the total DoE experiment is lessened.

An interesting method of visualizing the error of the lead-free experiment is shown in Fig. 2.2. The 95 percent confidence interval of the error, as measured by $2\sigma_{\text{error}}$ from the error variance, is shown superimposed on the graphical plot of the factors and levels. It can be seen that in factors that are not significant, the error span does not allow for distinguishing between the two levels of these factors. Given this pooling and significance information, the expected value, as shown in Table 2.8, should be calculated only from significant factors as follows:

EV = experiments average – contribution $(B1 + D2)$

EV = 16.94375 – [(16.95 – 6.1375) + (16.95 – 5.0125)] \
= –5.7938 defects/PCB or defect-free

Factor	Pooled	DOF	SS	MS	F	SS'	% Contribution
Finish	y	—	—	—	—	—	—
Paste	n	1	1868.401	1868.401	4.652	1466.748	15.66
F × P	y	—	—	—	—	—	—
Atmosphere	n	1	2277.676	2277.676	5.671	1876.023	20.03
F × A	y	—	—	—	—	—	—
P × A	y	—	—	—	—	—	—
F*P*A	y	—	—	—	—	—	—
Replication error		8	1916.505	239.563			
Pooled error		5	3304.978	660.996			
Total error		13	5221.483	401.653		6024.788	64.32
Total		15	9367.56	624.504		9367.559	100.00

TABLE 2.11 Lead-free Solder Experiment Analysis with Pooled Error

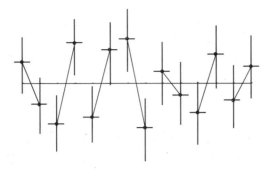

FIGURE 2.2 Visualizing the error of the lead-free experiment.

The negative value is indicative of the large experiment error (64 percent), hence the large ±CI in the EV.

2.5.4 Using DoE Methods for Various Phases of Green Material and Process Selection

As mentioned in Chap. 1, the search for new green materials and processes can take on various phases depending on the technology and maturity of the green materials. When a material is considered hazardous and is mandated to be banned or replaced by a green substitute, the material suppliers and company researchers will rush to recommend many different alternatives and tout their superior properties. DoE is an excellent tool to evaluate these green alternatives.

Partial Factorial DoE for Feasibility Study of Green Material Process and Material Selection

During the initial phases of a green material project, the tendency is to study many types of materials and their processing methods. Partial factorial DoEs are best suited to that task. An example would be the lead-free project phase I, discussed briefly in Chap. 1. The team wanted to study the feasibility of lead-free soldering and ensure that it is possible to obtain visual defect-free soldering and equal or better reliability than tin lead-based soldering. The design of the experiment proceeded as follows:

- *Design of experiments array:* One large experiment L27 was chosen for materials and processes which provides for three primary factors at three levels. These factors were assigned to two material selections (solder and surface finish) and one processing factor (time above liquidus, or TAL). Two other processing factors (reflow soak profile and atmosphere of nitrogen, yes or no) were assigned to columns which confound with three-way interaction. If two level factors were selected, then the presumed dominant level according to the team experience is repeated 2 times, for a total of three levels, such as two levels of OSP versus one level of ENIG surface finish factor, or two levels of nitrogen versus one level of air in the atmosphere factor.

- *Material selection decisions:* Three solder alloy materials were chosen, tin/silver (Sn/Ag), tin/silver/copper (Sn/Ag/Cu) and tin/bismuth (Sn/Bi) which were the leading candidates at that time. Two surface finishes—*organic surface passivation* (OSP) and *electroless nickel immersion gold* (ENIG) were selected for study. HASL finish was not considered because of problems with nonuniformity, and no other immersion finishes were considered such as silver finish since they were deemed too expensive. Tinned and palladium component finishes were used, but not part of the factor selection. The mix would represent a typical one found in PCBs.

- *Solder reflow processing parameter decisions:* TAL was considered important to minimize the thermal shock to the electronic components, and levels of 60, 90, and 120 seconds were considered. Since it is not easy to set up a reflow oven to implement the three levels, the design team combined the temperature ramp, the TAL, and the maximum temperature reached under TAL into one factor of 9 profiles, repeated 3 times. A discussion of that brainstorming technique was given in Sec. 2.5.2 under "Steps in Conducting a Successful DoE." Two thermal ramp profiles were considered, one linear and another called soak profile, which allowed for a preheat period. The linear profile was considered dominant and it was selected as the third level in the L27.

- *Nitrogen environment:* The use of a nitrogen environment was considered by the industry as beneficial to the soldering process. So two levels were chosen, nitrogen or air, with nitrogen considered dominant and used as the third level.

- Tests to be performed and recorded as the quality characteristics:
  1. Visual inspection based on electronics industry standards (IPC)
  2. Reliability simulation by pulling the component leads from a selected SOIC after 2000 h of temperature cycling.

In addition, six sets of two PCBs each were made of legacy solder (tin lead) and tested under the same visual and pull tests to act as a baseline. Only the material factors were considered for solder alloy (three levels given above) and surface finish (two levels given above). Processing and thermal reflow of the PCPs were performed at the commonly recommended historical levels.

The experiment array design and results obtained are shown in Table 2.12 for visual tests. The results were obtained from measuring

Sl. no.	Paste	Surface Finish	TAL, s	Soak	Nitrogen	Board Label		Profile No.	Board Faults Visual		Total	Average
1	Sn/Ag/Cu	OSP	60	Yes	Yes	1A	1B	1	797	944	1741	870.5
2	Sn/Ag/Cu	OSP	90	No	No	2A	2B	8	1213	1146	2359	1179.5
3	Sn/Ag/Cu	OSP	120	No	Yes	3A	3B	6	874	890	1764	882
4	Sn/Ag/Cu	ENIG	60	No	No	4A	4B	7	544	594	1138	569
5	Sn/Ag/Cu	ENIG	90	No	Yes	5A	5B	5	0	0	0	0
6	Sn/Ag/Cu	ENIG	120	Yes	Yes	6A	6B	3	0	0	0	0
7	Sn/Ag/Cu	OSP	60	No	Yes	7A	7B	4	828	819	1647	823.5
8	Sn/Ag/Cu	OSP	90	Yes	Yes	8A	8B	2	902	960	1862	931
9	Sn/Ag/Cu	OSP	120	No	No	9A	9B	9	1182	1164	2346	1173
10	Sn/Bi	OSP	60	No	Yes	10A	10B	13	1134	963	2097	1048.5
11	Sn/Bi	OSP	90	No	Yes	11A	11B	14	875	1136	2011	1005.5
12	Sn/Bi	OSP	120	Yes	No	12A	12B	12	967	1146	2113	1056.5
13	Sn/Bi	ENIG	60	No	Yes	13A	13B	13	1024	960	1984	992
14	Sn/Bi	ENIG	90	Yes	No	14A	14B	11	1016	1002	2018	1009
15	Sn/Bi	ENIG	120	No	Yes	15A	15B	15	843	560	1403	701.5
16	Sn/Bi	OSP	60	Yes	No	16A	16B	10	1148	1067	2215	1107.5
17	Sn/Bi	OSP	90	No	Yes	17A	17B	14	781	606	1387	693.5

TABLE 2.12 Phase I L27 Array and Visual Defects Results

Sl. no.	Paste	Surface Finish	TAL, s	Soak	Nitrogen	Board Label		Profile No.	Board Faults Visual		Total	Average
18	Sn/Bi	OSP	120	No	Yes	18A	18B	15	765	882	1647	823.5
19	Sn/Ag	OSP	60	No	No	19A	19B	7	1212	1279	2491	1245.5
20	Sn/Ag	OSP	90	Yes	Yes	20A	20B	2	1131	988	2119	1059.5
21	Sn/Ag	OSP	120	No	Yes	21A	21B	6	1027	933	1960	980
22	Sn/Ag	ENIG	60	Yes	Yes	22A	22B	1	0	0	0	0
23	Sn/Ag	ENIG	90	No	Yes	23A	23B	5	0	0	0	0
24	Sn/Ag	ENIG	120	No	No	24A	24B	9	180	240	420	210
25	Sn/Ag	OSP	60	No	Yes	25A	25B	4	796	829	1625	812.5
26	Sn/Ag	OSP	90	No	No	26A	26B	8	1205	1146	2351	1175.5
27	Sn/Ag	OSP	120	Yes	Yes	27A	27B	3	868	935	1803	901.5

TABLE 2.12 Phase I L27 Array and Visual Defects Results (*Continued*)

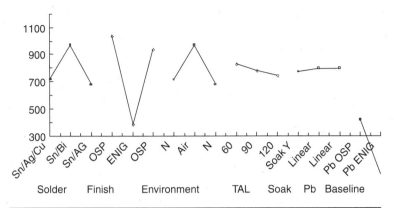

FIGURE 2.3 Phase I Lead Free Graphical Analysis for Visual Defects.

the visual defects for each of the 54 PCBs, with a defect opportunity of 1279. The visual defect graphical analysis is given in Fig. 2.3. It interesting to note that the levels that were duplicated such as nitrogen and linear soak showed similar visual defects, which is a good indicator of a good error-free experiment. Baseline lead defects showed lower numbers.

To the delight of the experiment design team, there were several instances of zero visual defects, as shown in Table 2.12, showing that lead-free (green) electronics packaging such as PCBs and component finishes was indeed a possibility, especially since this experiment was performed in 1999/2000, six years before mandated regulation (RoHS) was proposed for green electronics products.

To evaluate the green material and process alternative, an ANOVA was performed. The following is a conclusion of the ANOVA data, as well as the graphical data analysis in Fig. 2.3. In addition, the pull tests for the visual zero-defects PCBs were analyzed in Tables 2.3 and 2.4. The conclusions were summarized for visual defects in Table 2.13.

- Metallurgy of solder paste alloy and surface finish is the most critical factor in producing visually good solder joints, with the surface finish being the biggest contributor (53 percent) and the solder composition alloys contributing 7 percent.

- Nitrogen reflow environment improves the visual quality of solder joints having 10 percent statistical contribution.

- The thermal profile of solder reflow is not significant for visual defects, including the amount of TAL and thermal profile.

- Optimal settings that can result in zero defects with careful process control:
  - Solder composition of Sn/Ag or Sn/Ag/Cu alloys.
  - ENIG surface finish is superior to OSP.
  - Nitrogen reflow environment is recommended.

Factors	Significant	Contribution to the Experiment	Level	Average Defect Count	Tin/Lead Baseline
Solder paste alloy	Yes	7.16%	1. Sn/Ag/Cu	714	217.75
			2. Sn/Bi	937	
			3. Sn/Ag	709	
PWB surface finish	Yes	52.71%	1. OSP	987	420
			2. ENIG	386	15
Time above liquidus	No	0.57%	1. 60 s	829	
			2. 90 s	783	
			3. 120 s	747	
Soak	No		1. Soak	770	
			2. Linear	795	
Nitrogen	Yes	10.72%	1. Nitrogen	695	80
			2. Air	969	355
Average defects				787	217.75
Total defect opportunity				1279	

TABLE 2.13 Phase I L27 Array Visual Defects Analysis Conclusions

- When setting the materials and process to the preceding alternatives, zero visual defects could be obtained, and the reliability as measured by pull testing of some component leads after thermal cycling showed a higher average of force needed to break the solder joint, but with similar variability.

- Phase I showed the feasibility of zero defects and superior reliability of green lead-free materials and processes long before the deadline for mandated green directives such as RoHS.

Full Factorial DoE for New Green Material Interactions

As indicated in Chap. 1, once the green feasibility study is completed, further work is required to study how this green material will interact with a wider selection of current green materials in quality as well as reliability, which is called phase 2 in Chap. 1. Phase 3 (called TURI 3 in Chap. 6) is the final confirming test of the green materials and involves the optimization of the manufacturing process and finding out other issues such as repair and rework of green materials.

Full factorial DoE design is usually required after an initial investigation such as the one in phase 1, since the focus is shifted to performance of the material in high production volume and its interaction with existing materials and processes. The DoE design space is fixed, and the green project team should attempt to optimize the green material and process within that space. Phase 2 was a full factorial DoE into the processing parameters and compatibility of the lead-free solders with component types and finishes.

Tests performed in phase 2 included multiple t tests for different integrated circuit (IC) types, such as SOIC and QFP. A total of 120 multiple range tests were performed for each type, comparing the effects of solder suppliers, surface finishes, component finishes, before and after thermal cycling, and lead-free versus tin lead baseline for visual and pull tests. Of all these tests, only a few showed significant differences, including the set of OSP, SAC solder supplier B and tin/lead component finishes. This was apparent to electronics assembly suppliers, since they recommend special consideration when mixing leaded and lead-free component finishes (see the Chap. 6 from Benchmark Corporation, where this phenomenon is called hybrid soldering).

In conclusion, phase 2 of the lead-free project showed that careful analysis and testing are needed when specifying PCB surface and component finishes with SAC (green) solders. However, lead-free soldering can be accomplished with a large variety of materials and processes, and from multiple suppliers.

The DoE for phase 3 was a full factorial design of five factors with two and three levels: solder supplier (two levels), laminate surface finish (three levels), laminate type (two levels), component lead finish (two levels), and pull direction (two levels). The last factor was added to ensure the integrity of the component lead pull process, to see if the pull force was consistent from front to back. The pull force interaction plot in Fig. 1.2 of laminate finish versus laminate type indicates that laminate supplier B interacts differently with the three laminate finishes (ENIG, immersion silver, and OSP), while laminate supplier A has the same pull force result with any of the three laminate finishes selected. This is an important result when one is investigating technology and material property claims by the green material suppliers.

2.5.5 DoE Section Conclusions

It has been shown, through several examples, that DoE is an excellent tool to investigate and evaluate new trends and technologies in green material development and properties. There are several techniques in DoE that should be thought out well in advance: the definition of the characteristics to be optimized, the selection of factors and levels, the treatment of factor interactions, the selection of experiment arrays, and how to simulate and measure variability and error.

   An initial DoE project should be selected carefully to optimize a green design that is being developed for the first time, with many factors and levels, preferably in a partial factorial DoE. Knowledge gained from the first green DoE should be used to further understand the material properties and green process parameters by follow-on DoEs, using the full factorial methodology to ensure the study of all relevant interactions of materials and processes.

References and Bibliography

Box, G., S. Bisgaard, and C. Fung. *An Explanation and Critique of Taguchi's Contribution to Quality Improvement.* University of Wisconsin. 1987.
Cochran, W., and G. Cox. *Experimental Designs,* 2nd ed. New York: Wiley, 1981.
Diamond, W. J. *Practical Experiment Design.* New York: Van Nostrand Reinhold, 1981.
Ealy, L. "Taguchi Basics." *Quality Journal,* November 1988, pp. 26-30.
Guenther, W. *Concepts of Statistical Interference.* New York: McGraw-Hill, 1973.
Hicks, C. *Fundamental Concepts in the Design of Experiments.* New York: McGraw-Hill, 1964.
John, P. *Statistical Analysis and Design of Experiments.* New York: Macmillan, 1971.
Johnson, R. *Probability and Statistics for Engineer.* 5th ed. Englewood Cliffs, N.J.: Prentice-Hall, 1994.
Phadke, M. *Quality Engineering Using Robust Design.* Englewood Cliffs, N.J.: Prentice-Hall. 1989.
Ross, P. *Taguchi Techniques for Quality Engineering.* New York: McGraw-Hill, 1987.
Roy, R. *A Primer on the Taguchi Method.* New York: Van Nostrand Reinhold, 1990.
Shina, S. *Six Sigma for Electronics Design and Manufacturing.* New York: McGraw-Hill, 2002.
Shina, S. "Design of Experiments," Chapter 25 in *Environment Friendly Electronics: Lead-Free Technology* by J. Hwang. New York: Electrochemical Publications, 2001.
Shina, S. *Successful Implementation of Concurrent Engineering Products and Processes.* New York: Wiley, 1994.

Shina, S. *Concurrent Engineering and Design for Manufacture of Electronic Products.* New York: Van Nostrand Reinhold, 1991.

Taguchi, G. *Introduction to Quality Engineering.* Tokyo: Asian Productivity Institute, 1986.

Taguchi, G. *System of Experimental Design.* White Planes, N.Y.: UNIPUB-Kraus, 1976.

Taguchi, G., El Sayed, E., and Hsiang, T. *Quality Engineering in Production Systems.* New York: McGraw-Hill, 1988.

Young, H. *Statistical Treatment of Experimental Data.* New York: McGraw-Hill, 1962.

Walpole, R., and R. Myers. *Probability and Statistics for Engineers and Scientists.* New York: Macmillan, 1993.

CHAPTER 3

Reliability of Green Electronic Systems

David Pinsky
Raytheon Integrated Defense Systems, Tewksbury, Massachusetts

This chapter details how the transition to green electronics is affecting reliability. The introduction describes common measures of reliability and how they relate to system-level requirements. A detailed discussion follows covering the response of lead-free solder joints to environmental stress, and how this compares with the legacy tin/lead systems. The effects of rework and repair on reliability are covered next. The complex topic of tin whisker risk, assessment, and mitigation techniques and strategies is presented. Case studies are provided in novel failure modes associated with green materials and processes.

3.1 Introduction

3.1.1 What Is Reliability?

One common, simple definition of *reliability* is quality as a function of time, or the ability of a system to continue to perform as intended after suffering the vicissitudes of usage, transportation, aging, and environmental exposure.

The reliability of the system must always be considered within the context of the requirements of the application and the environments to which the system is exposed. A cellular telephone that functions properly for two years and then suddenly ceases to operate altogether may be deemed to have performed in a "reliable" manner, whereas a cardiac pacemaker that begins to degrade in performance after four years may be deemed to have performed in an "unreliable" manner. It is important always to keep this in mind when considering the details of any system within the context of reliability. Solder joints, circuit boards, and the other constituents of complex electronic systems will exhibit a range of characteristics, none of which automatically imply reliable or unreliable performance. The specific characteristics must be considered to determine their affect on the system and in this effect must be compared against an expectation before meaningful value judgments can be made about reliability.

3.1.2   Measures of Reliability

The details of the mathematics of reliability are beyond the scope of this chapter, and have been amply discussed elsewhere. However, a brief discussion is offered here to provide some context for the reliability data presented below.

Reliability of systems and their constituents is quantified in a number of ways. The most common form of metric involves defining the relationship between the duration of exposure to a given environment and the occurrence of failure events. Results are typically expressed in the language of probability and statistics.

It is standard practice for quantifying reliability to assume that there is a well-behaved probability density function that describes the probability of failure as a function of time. Once real reliability data are available, the validity of this assumption is then tested. In general, *time* implies the duration of exposure to an environment of interest, and *failure* implies degradation of a performance parameter of interest. If one integrates the probability of failure over time, the resulting function is the cumulative number of failures as a function of time. The complementary function, which is the integral of the probability function from a given time to infinity, gives the number of units that have not yet failed at a given time and is termed the *survival function*.

In practice, reliability in a data set is collected through either experiment or field usage, and then the data set is fit to a probability distribution. Once the data have been fit to a given function, predictions can be made based on the assumption that this function adequately describes the performance of the system. These results can then be extended to compare the system with other systems that have been analyzed similarly. The most common probability distribution that is used to fit failure probability data is the Weibull distribution. However, it is good practice to consider a variety of distribution functions to check to see which yields the best fit.

Reliability metrics can be either measured directly or calculated based upon determining a characteristic probability distribution as discussed above. Two of the most common metrics that are used to quantify reliability are as follows:

1. *Mean time between failures (MTBF).* This metric is most commonly used to describe the performance of higher-level systems, where repair and maintenance activities are

commonly performed. In such applications the expectation is that some failures will be inevitable over the lifetime of the system. What is important is that such failures do not occur with great frequency. As a system ages and the probability of failure during a given period of time increases, the MTBF is expected to decrease. This is a phenomenon that should be quite familiar to anyone who has owned an automobile over an extended period of time.

2. *Time to x percent failure N_x.* The metric N_{50} describes the median lifetime of the population. This provides a rough estimate of how long an individual member or item of the population should be anticipated to survive. Sometimes the metric N_{63} is used in a similar fashion as an indicator of lifetime.

   Under circumstances where system performance is critical, or multiple items must all be functional, the time to reach much smaller percentages will be of interest. In such cases the time to achieve cumulative failure rates of 1 percent in the parts per thousand, parts per million, and even parts per billion may be used.

3.1.3 Weibull Distribution Example

The most commonly used method of comparing life test data, and of using them to predict service life, involves fitting the data to a probability distribution. The most common probability distribution used for this purpose is the Weibull distribution. However, best statistical practices dictate that the distribution to be used should be the one that provides the best fit for the data.

Since the Weibull distribution is so common, an example is provided here of how life test data are fit to a Weibull distribution and how different life test data sets can be compared to one another and used to make reliability predictions.

The two-parameter Weibull probability distribution function is defined by the following equation.

$$f(T) = \frac{\beta}{\eta}\left(\frac{T}{\eta}\right)^{\beta-1} e^{-\left(\frac{T}{\eta}\right)^{\beta}} \tag{3.1}$$

where $f(T) \geq 0$, $T \geq 0$, and β, $\eta > 0$.

The two parameters that define a particular Weibull distribution are the shape parameter indicated by β and the scale parameter indicated by η. This particular probability distribution function gained popularity for performing reliability predictions during the middle of the Twentieth Century because it often provided a good fit for a wide range of lifetime data, and was amenable to graphical integration techniques. Its popularity has not decreased since the advent of computers that have rendered the graphical integration techniques obsolete.

It is common practice to utilize special-purpose reliability software to perform data fitting and attendant calculations. All the plots and fits provided in the examples below were prepared using one such popular commercially available software package — ReliaSoft's Weibull++.

Example 3.1 Weibull Distribution   To begin, let us assume that life testing has been performed on a population of 12 packages that were soldered to a printed-circuit board (PCB), that testing was performed until all 12 failed, and that the time to failure was recorded for each. The concept of time in this example is equally applicable to the number of thermal cycles, duration of exposure to vibration, or length of time exposed to any particular test environment.

The hypothetical lifetimes are 473, 826, 947, 1252, 1389, 1746, 2012, 2046, 2243, 2684, 2903, and 3500.

The data set was then loaded into the software for fitting to a two-parameter Weibull distribution, and the fit shown in Fig. 3.1 was obtained:

Note that the Weibull function, when plotted on this particular scale, defines a straight line with the slope determined by β and the intercept determined by η. Several commonly employed metrics can be easily derived from this plot including the median lifetime N_{50} and the time to 1 percent cumulative failure

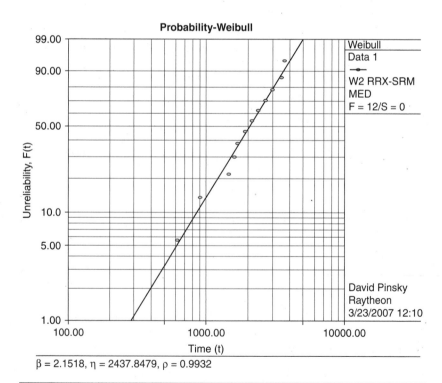

$\beta = 2.1518$, $\eta = 2437.8479$, $\rho = 0.9932$

Figure 3.1   Hypothetical life test data set 1 fitted to a Wiebull distribution.

$N_{0.01}$. Depending on how you need to use the data, one or the other of these metrics may prove more interesting. If you are planning spares management, and you want to know how many spares to stock and how fast you must supply them, then knowing when one-half of the parts in the field will have failed will give you an good estimate for how many parts you will need and how soon. If, however, you are looking at ensuring reliability of the individual system to very high level, and the system contains a number of the items you are looking at, then the 1 percent number may be more meaningful.

Once the Weibull fit has been performed, the data can be looked at in different ways, which are typically supported by the special-purpose software. Commonly used plots include unreliability as a function of time, reliability as a function of time, failure rate as a function of time, and the failure probability density function.

This single set of data and the fit can be used to make life predictions for other items that are identical to the items that have been tested. It is a simple matter of applying the probability distribution for the parameter of interest. When performing tests for the purpose of making life predictions, it is strongly recommended to test samples to failure. Although predictions can be made using data on samples that have not failed by a given time (suspended testing), it is good practice to continue testing until at least 63 percent of the population has been taken to failure. Otherwise, a large degree of uncertainty will exist in the lifetime predictions.

Example 3.2 Comparing Data Sets of the Weibull Distribution   Let us now consider a second set of life data. For this example assume that it represents a second life test performed on 12 packages soldered to a PCB, tested to the same environment as was used in Example 3.1. Again, testing has been performed to failure, and the time to failure was recorded in every instance. However, the second set of components utilized a different solder material than did the first set. The lifetimes for data set 2 are 372, 986, 1437, 1686, 2092, 2543, 2761, 2802, 3312, 4118, 4692, and 6841.

These data can be fitted distribution and then can be used to make life predictions for this material, just as in Example 3.1. In addition, it is possible to compare the two data sets and to gain insight into their relative performance. For this purpose both data sets are plotted on the same chart in Fig. 3.2.

It is interesting to note that the two lines intersect. This is so because they exhibit different values of β, the shape parameter or slope. This creates a complex situation when you are attempting to determine which material provides "better" reliability. If your principal concern is how long it takes for one-half of the parts to fail, you compare the relative values of N_{50} and can see that material set 2 is "better" by this definition. However, if your principal concern is when you experience early failures, by comparison of the relative values of $N_{0.01}$ it can be seen that material set 1 is preferred.

Example 3.2 illustrates that caution must be used when making decisions based upon reported data from life testing. It is very common for researchers to report various values of N_x when reporting abbreviated summaries of life testing. These numbers are directly comparable only when the β values are fairly close for all the distributions involved. Otherwise, the relative ranking of the various options when viewed from the perspective of a potential user may be different from that reported in the short summary.

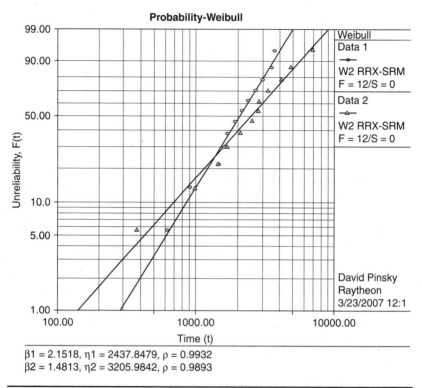

β1 = 2.1518, η1 = 2437.8479, ρ = 0.9932
β2 = 1.4813, η2 = 3205.9842, ρ = 0.9893

Figure 3.2   Hypothetical life test data sets 1 and 2 fitted to Wiebull distributions.

3.1.4   Reliability in the Context of Green Electronics

The reliability of electronic systems is a very complex topic, which has been studied and reported on for many decades. It is the intent here to focus on those aspects of the transition to green electronics that may have an impact on reliability.

The transition toward green electronics is resulting in changes in the set of materials that are used in electronic packaging and interconnections. The new set of materials will exhibit different physical properties from the legacy set. These differences will cause different responses to the applied environmental stress and therefore may result in differences in the system reliability.

Changes are also occurring in the set of manufacturing processes that are being used in fabrication and assembly of electronic systems. New processes may result in different initial product quality levels than were experienced previously. This may have a significant effect on reliability. In addition, different process conditions may adversely affect some of the materials that have not been altered during the transition to green electronics. This could also affect reliability. In this chapter we examine these changes to assess how they will affect system reliability.

The most common statement of the underlying question here is, Will the new green system be as reliable as the old legacy system? Such a question is difficult enough to answer if the legacy system and the new system are essentially identical. However, the world of electronic packaging is constantly moving forward. The result is that the new, green systems will be denser, higher-power, and more complex than the legacy systems, which they are supplanting. This greatly complicates the situation, because we are comparing against a moving baseline. The question should be, How will the reliability of my new green system compare to that of a new system manufactured using legacy materials?

3.2   Lead-Free Solder Interconnections

3.2.1   Lead-Free Solders and the Tin/Lead Baseline

Nearly all our historical experience with electronics reliability, and all our field experience with very long-term reliability of electronics, comes from systems fabricated using tin/lead solder. Therefore, if we wish to leverage off of this store of experience, we must first understand the properties of tin/lead relative to the various lead-free solders.

Tin/lead eutectic solder was "down-selected" as the solder of choice very early in the history of electronics manufacturing principally due to its superior processing properties. Once the industry had established tin/lead as the standard alloy for assembly, all the other materials and processes evolved for compatibility with tin/lead processing.

The tin/lead binary system forms a simple eutectic. One important property of a eutectic alloy is that, at eutectic composition, it melts and freezes at a single temperature, called the *eutectic temperature*. For the tin/lead system this is 183⁰C. Alloys that solidify at a single temperature will typically exhibit a uniform microstructure and the smooth external surface.

If the alloy exhibits a composition different from the eutectic composition, the transition between solid and liquid occurs over a range of temperatures, sometimes referred to as the "mushy zone" or "pasty range." Alloys that freeze over a range of temperatures will typically solidify by dendritic growth, resulting in a complex multiphase microstructure. In addition, interdendritic shrinkage will result in a rough surface and potentially small fissures at the surface. In general, smooth surfaces and uniform microstructure yield mechanical properties superior to those obtained from rough surfaces and nonuniform microstructure.

Another important property of solder alloys is the temperature at which they will reflow. This drives the temperature to which assemblies must be subjected during manufacturing. It also constrains the temperature to which the assemblies may be safely subjected after manufacture. Clearly, lower processing temperatures result in less expensive processes and reduced stress on components during assembly. However, if the processing temperature is too low, there may be a risk that this temperature could be approached during high-temperature operating excursions, with severe consequences for reliability. The tin/lead eutectic temperature of 183^0C strikes a nice balance between these two constraints. Solder alloys that reflow at temperatures higher than this will induce greater stress in the assembly during manufacturing (and especially during rework, as discussed below). Solder alloys that reflow at temperatures lower than this will be at greater risk for degradation during high-temperature operation. Table 3.1 compares the melting and freezing temperatures of tin/lead with those of several important lead-free solder alloys.

The last processing parameter of concern for solder alloys may be the most important: wetting behavior. Wetting describes the ability of the liquid to spread out and cover, or "wet", a solid surface. Wetting is defined and measured by various means including the wetting angle, time to wet, and wetting balance force. Eutectic tin/lead offers superior wetting performance for all standard solderable finishes. This is not surprising, because these finishes were selected originally for compatibility with tin/lead. Also, tin/lead was originally selected in large part because of its superior wetting ability. As a rule, none of the lead-free alloys exhibit the same degree of excellent wetting exhibited by tin/lead. The net result is that lead-free solders will not spread and flow to the same degree that tin/lead solders will. This in turn will affect the shape of the resultant solder joints.

Alloy	Melt/Freeze Range, °C
Sn63 Pb37	183
Sn62 Pb36 Ag2	177–189
SAC	217–220
Sn96.5 Ag3.5	221
Sn99.3 Cu0.7	227
Sn42 Bi58	138
Sn95 Sb5	232–240
Sn48 In52	118

TABLE 3.1 Melt and Freeze Temperatures of Various Solder Alloys

3.2.2   Lead-Free Solders and Their Properties That Affect Reliability

Solder alloys exhibit a range of properties, some of which affect reliability more strongly than others. Table 3.2 lists a range of important mechanical properties of these solders. It is important to note that published data vary widely. The values in this table are presented primarily for comparison purposes. Values will depend upon temperature, strain rate, grain size, and compositional variations. These values are for room temperature and at moderate strain rates between 0.2 and 5.0 mm/s.

The mechanical properties of solders are critical in determining the long-term behavior of soldered assemblies. There are meaningful differences between the mechanical properties of the tin/lead baseline and the lead-free alternatives. The mechanical behavior of solder alloys is greatly complicated by the fact that these alloys are used at temperatures that represent a very high fraction of their absolute melting point and always above the minimum recrystallization temperature. Therefore, these alloys are being used at "high temperature." The result is that these alloys will respond to applied stress viscoplastically. Viscoplastic deformation means that stress depends not only upon strain, but also on the strain rate. The best analog for viscoplastic behavior from common experience is Silly Putty. If you pull on it very fast, it will deform elastically and fracture at a relatively high stress. If you pull on it very slowly, it will stretch to a much higher elongation and eventually fracture at a much lower stress. This effect is illustrated schematically in Fig. 3.3.

Many of the important stress conditions encountered during operation of electronic assemblies involve moderate stress that is applied over long periods of time, the exact conditions where viscoplastic behavior is important. Therefore, when comparing the mechanical behavior of various solder alloys, consideration of short-duration tensile properties is of limited utility.

Solder Alloy	Tensile Strength, MPa	Young's Modulus, GPa
Sn63 Pb37	40–45	29
Sn62 Pb36 Ag2	45–60	23
SAC	50–70	45–50
Sn96.5 Ag3.5	30–40	43
Sn99.3 Cu0.7	25–35	

TABLE **3.2**   Mechanical Properties of Selected Solders

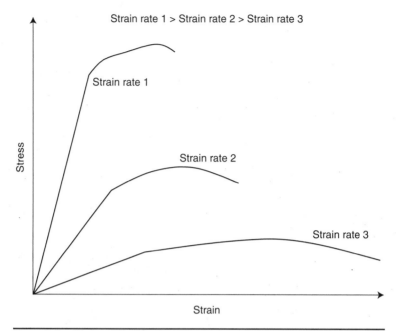

Strain rate 1 > Strain rate 2 > Strain rate 3

Strain rate 1

Stress

Strain rate 2

Strain rate 3

Strain

Figure 3.3   Illustration of viscoplasticity.

Another consequence of the fact that solder alloys operate at high temperature is that the internal microstructure and grain size can change over time, a process referred to as *aging*. Changes in the internal microstructure grain size can have important effects on the macroscopic mechanical properties. This phenomenon has been well documented for tin/lead systems.

These two time- and temperature-related processes, viscoplasticity and aging, must always be considered when attempting to interpret and understand test results and how they compare with use environments.

One peculiar challenge associated with these time-dependent properties is that meaningful testing requires extended periods of time to perform. This extended test duration results in a fairly long cycle time between alloy development and measurement of critical mechanical behaviors. This is particularly problematic when one considers that the set of lead-free alloy compositions currently in use is far from stable. We are still early in the implementation of lead-free electronic manufacturing, and solder alloy variations are still being implemented to solve a variety of problems as they arise. This has resulted in an out-of-phase condition between the development of long-term reliability data and the implementation of various solder alloys in manufacturing.

3.2.3   Test Environments of Greatest Interest

Past experience with tin/lead soldered systems indicates that the most important considerations for solder-joint reliability have been the initial joint quality, imposed thermal cycling environment, and imposed mechanical shock and vibration environment. Direct corrosion of solder joints has rarely resulted in unreliability of electronic systems, and it should also be true for most lead-free solder systems under consideration (with the possible exception of solders containing zinc, which may be more susceptible to corrosion than the legacy tin/lead alloy).

Thermal Cycling

Thermal cycling is the environment that has received the greatest amount of attention when one is considering the reliability of soldered interconnections.

The principal sources of thermal cycling stress of electronic assemblies are internal power dissipation and external thermal transients. The magnitude and time dependence of the thermal environment for any particular assembly will be a combination of these two stresses. Electronics equipment that is required to operate outdoors without the benefit of environmental protection is generally assumed to be subjected to one thermal cycle per day. Avionics systems that operate in non-environmentally controlled conditions will be subjected to one cycle per flight segment (commercial aircraft) or sortie (military aircraft). Satellite electronics will be subjected to one thermal cycle per orbit. Some typical externally induced temperatures are listed in Table 3.3.

Internal power dissipation will typically result in a fairly repeatable temperature increase of the assembly with each power cycle. The frequency of the power cycling will be determined by the operation of the system in question. There are many network and server applications that involve 24/7 operation and therefore experience minimal power cycling over their lifetimes. Laptop computers and other portable devices are more likely to see multiple power cycles over the course of a typical day.

Thermal cycling induces strain into solder joints as a result of differential thermal expansion between the electronic components in PCBs. The magnitude of the displacement induced across an electronic component in the temperature changes between an isothermal state at temperature $T1$ and a second isothermal state at temperature $T2$ is illustrated in Fig. 3.4 and is given by Eq. (3.2).

$$\text{Displacement} = X\ \Delta T\left(\text{CTE}_{\text{component}} - \text{CTE}_{\text{pwb}}\right) \qquad (3.2)$$

The induced displacement is proportional to the size of the component and also to the magnitude of the temperature difference. Equation (3.2) predicts the displacement between two isothermal

Use Environment	Nonoperating Temperatures, °C	Operating Temperatures, °C
Office/home	−40 – +85	+10 – +30
Data center	−40 – +85	15 – +75
Ground-based communications (uncontrolled environment)	−40 – + 85	−40 – +85
Industrial	−55 – +125	−40 – +85
Automotive	−40 – +100	−40 – +85
Aircraft (controlled temperatures)	−40 – +65	0 – +55
Aircraft (uncontrolled temperatures)	−55 – +125	−40 – +100
Satellite	−65 – +125	−65 – +125

TABLE **3.3** Temperature Extremes Associated with Various Common Applications

states. Additional displacement can be induced as a result of the presence of thermal gradients within an assembly. If the source of heating and cooling is external to the assembly, these gradients will typically be small and are unlikely to be strong contributors to the overall induced displacement. However, if one portion of electronic assembly dissipates significant levels of power, thermal gradients within the assembly can be large, and the displacements induced by this nonisothermal situation may be significant and should not be ignored.

Displacement between electronic components and PCBs will induce strain into the solder joints that interconnect them. The amount of strain induced will depend upon the geometry involved and upon the amount of mechanical compliance present in the body of the component, the component leads, and the PCB itself. Stiffer mechanical elements will result in higher levels of displacement driven into the solder interconnections, and higher levels of strain.

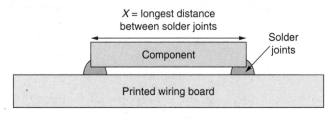

FIGURE **3.4** Displacement resulting from thermal cycling.

Cyclic loading of mechanical members will typically result in failures at loads below which fracture would have occurred on the first cycle. (All paper clip benders are well acquainted with this effect.) This phenomenon is referred to as *fatigue*. Fatigue life behavior of structural metals has been studied extensively since the middle of the 19th century, and many of the resulting models and theories have some applicability to solder interconnections.

Fatigue is typically divided into two regimes: *high-cycle* fatigue and *low-cycle* fatigue. Low-cycle fatigue usually involves strains approaching or even exceeding yield, with resulting cycles to failure of approximately 10^4 or fewer. High-cycle fatigue involves strains well below yield, and greater than approximately 10^4 cycles to failure. Life-limiting thermal cycling for PCBs typically involves fewer than 10^4 cycles and therefore falls within the realm of low-cycle fatigue. Life-limiting vibration of PCBs typically involves greater than 10^6 cycles and therefore falls within the realm of high-cycle fatigue.

The most common presentation of fatigue life data involves the *S/N* curve, which plots induced strain versus number of cycles to failure for a population of identical test specimens. A generic example of such a curve is shown in Fig. 3.5.

When the number of cycles to failure is plotted logarithmically, significant portions of most fatigue life curves can be approximated by a straight-line fit. Such straight lines define a power-law relationship with respect to applied strain, with the strain exponent determined by the slope of the line. Various forms of these models have been developed and are used to predict solder joint life under both thermal cycling stress and vibration stress. When the strain is assumed to be entirely due to the thermal cycling, the models include

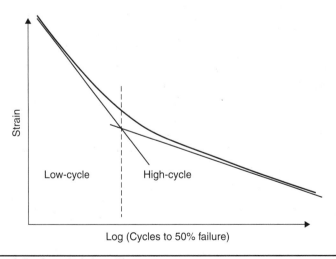

FIGURE 3.5   Generic fatigue life curve.

the term $T_{hot} - T_{cold}$ or ΔT in lieu of strain, based upon the assumption that strain relates linearly to the temperature change by a constant coefficient of thermal expansion. Models that are used to predict life under vibration typically express strain ε explicitly.

Relatively slow thermal cycling of solder joints results in an additional set of complications resulting from both viscoplastic effects and the temperature dependence on the mechanical properties of solder. For this reason, many thermal cycling fatigue life models for solder include terms that relate to the length of the thermal cycle and the absolute temperature over which the cycles are run.

So far all this discussion applies equally for tin/lead solder and for lead-free solders. The differences begin when we consider the detailed response of the solder joints to the induced displacements and the resulting fatigue response. Let us consider how tin/lead solder joints will respond in comparison with SAC 305 solder joints.

3.2.4 Thermal Cycling Fatigue Life Behavior of Tin/Lead and SAC Alloys

Thermal cycling of printed-wiring assemblies is the most common technique for assessing long-term reliability of the solder interconnections between the components and the PCB. Industry standard IPC-9701A, "Performance Test Methods and Qualification Requirements for Surface Mount Solder Attachments," is widely used to define the test conditions for these thermal cycling evaluations.

The preferred method requires that special "daisy-chained" components be mounted to a special-purpose PCB, which has been configured to provide a single continuous circuit path for each component. Each circuit path runs from the board through every solder joint of one component, by means of the internal daisy chain and matching board circuitry. Such a configuration is illustrated in Fig. 3.6.

These test boards are then subjected to thermal cycles that are comprised of a specified dwell time at each extreme, with thermal "ramps" between them. This is in contrast to thermal shock testing, which involves transfer of units under test directly between a hot environment and a cold environment. The comparison between thermal cycling and thermal shock testing is illustrated in Fig. 3.7, with cycling as the triangular and shock as the square waveform.

Continuity is measured for each component's circuit continuously throughout the duration of the test. An electronic event detector is used to record the precise time at which the initial indication of increased circuit resistance occurs for each component. The time of this event is then defined as the time to failure for that component. Note that each component is considered as a single datum, even though a single component might be attached to the PCB by any number of solder joints. The rationale for this convention is that the

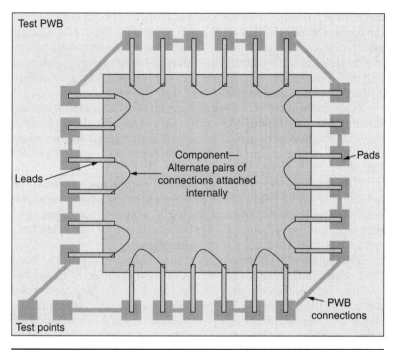

FIGURE 3.6   Daisy chain testing illustration.

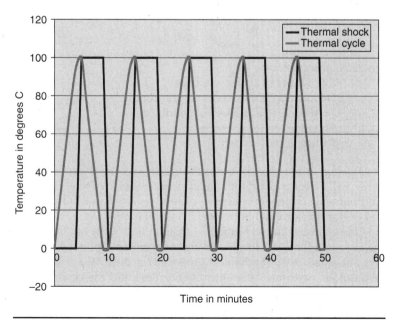

FIGURE 3.7   Thermal cycling and thermal shock.

mechanical stress in the solder joints is a function of the mechanics of the entire component and all the joints together. Failure of a single joint represents failure of this entire attachment system. Also, failure of a single joint implies functional failure of the component. Failure of additional solder joints for the component has no effect on reliability of the PCB and is therefore of no practical interest.

Typically, number of cycles to failure data are then fit to a statistical distribution, most commonly a Weibull curve. From the Weibull curve, it is common to extract metrics of characteristic life including median time to failure, time to 10 percent failure, and time to 1 percent failure.

Using data provided from the thermal cycling test to predict the reliability of actual field use requires the use of a model to transform reliability test results obtained during a relatively short test over relatively wide temperatures into the conditions of relatively long service life over relatively narrow temperatures.

Models have been developed and validated that provide good prediction of fatigue life of tin/lead solder joints subjected to thermal cycling. Such models can provide good agreement if they can predict the fatigue life of joints under test within a factor of 2. As mentioned above, a power-law relationship provides a reasonable estimation for a fatigue life curve. Therefore, the simplest model that is used to compare the fatigue life of solder joints exposed to different thermal cycling regimes is the power-law relationship (or simplified Coffin-Manson model) shown in Eq. (3.3).

$$\text{AF} = \frac{No}{N_t} = \left(\frac{\Delta To}{\Delta T_t} \right)^{\beta} \tag{3.3}$$

where AF = acceleration factor
N = number of cycles to failure
X_O = subscript indicates operation conditions
X_t = subscript indicates test conditions
ΔT = range of thermal cycle
β = fitting parameters

The higher the value of the exponent, the greater the acceleration factor will be for a given thermal cycling test relative to a given use environment. Exponent values typically reported are between 1.9 and 2.7 for tin/lead and between 1.5 and 3.0 for SAC.

One of the important findings is that, unlike for the tin/lead alloy, the exponent for SAC alloys is strongly dependent upon the package style. Because of the mechanical properties of the SAC solder joints, SAC solder performs differently for leadless and for ceramic-style devices. The net effect is that if you test the same specific PCB geometry manufactured using tin/lead and also with SAC alloys, under identical test conditions, the relative test severity will vary from component to component. When you use the above approach to predict the thermal cycling life of SAC alloy assemblies, it will be necessary to select the appropriate exponent. There is a significant difference between the results depending upon which exponent is selected. It is very likely that components that exhibit properties intermediate between the two cases that have been studied will be properly modeled by the use of exponent with an intermediate value. The conservative approach is to use the lower value of the exponent to cover all situations where the higher exponent cannot be justified explicitly. This can result in some degree of overtesting.

To illustrate this effect, consider two distinct cases: In the first instance, consider stiff components without leads, such as ceramic chip capacitors or ceramic ball grid arrays, where the solder joint itself will provide a significant portion of the overall mechanical compliance that must take up the imposed displacement. In the second case, consider components with compliant leads, such as a plastic encapsulated microcircuit with copper/alloy gull wing style leads, where the leads will provide most of the mechanical compliance, so that the load on the solder joint will be that which is required to move the leads a sufficient distance to account for the induced displacement.

Tin/lead has a lower modulus of elasticity and lower yield strength than does SAC 305. Therefore, in the first instance, where the displacement is fixed, significantly higher loads will be induced into the SAC 305 joints than would be introduced into identical tin/lead joints. In the second case, where the load is fixed, significantly higher strains will be induced into the tin/lead joints than will be introduced into the SAC 305 joints. These two cases for tin/lead solder joints compared with SAC 305 solder joints are illustrated in Fig. 3.8.

As mentioned above, the trend that has been identified to date is that SAC 305 outperforms tin/lead under low-strain thermal cycling conditions, but that tin/lead outperforms SAC 305 under high-strain thermal cycling conditions. Factors that lead to high-strain and to low-strain conditions are summarized in Table 3.4.

How does this behavior translate into assembly-level reliability? One useful visualization method is to consider the fatigue life curves for SAC 305 and for tin/lead with environmental stress plotted on the vertical scale. The two curves cross at some point. Below some critical environmental stress level, SAC 305 solder joints will last longer, while above this level tin/lead solder joints will last longer.

Exactly where this effective crossover point is for any given circumstance will depend upon the precise geometries involved. JP Clesch has reported the crossover at approximately 6 percent strain.

A simple power-law model does not account for effects relating to the duration of the thermal cycle or the maximum temperature achieved during the cycle.

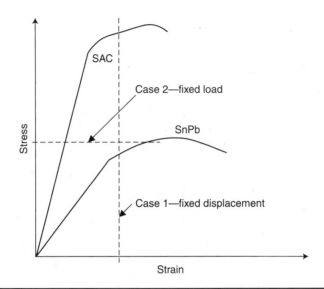

Figure 3.8   A comparison between the effects of high and low strain conditions for tin/lead and for SAC 305 solder alloys.

3.2.5   Effects of Thermal Dwell Time and Maximum Temperature Stress Relaxation or Creep

For tin/lead alloys, it was discovered that solder joints could survive a larger number of thermal shock cycles than they would survive slower thermal cycling iterations. This effect is due to the time and temperature dependence of the viscoplastic properties of solder.

High-strain contributors	Low-strain contributors
Large temperature excursions	Small temperature excursions
High levels of vibration	Little or no vibration
High levels of mechanical shock	Little or no mechanical shock
Large differences in the CTE between package and printed-wiring board	Close match of CTE between package and printed-wiring board
Use of packages without leads	Use of packages with compliance leads
Large amounts of board flexure during handling	Limited amount of board flexure during handling

Table 3.4   Common Factors That Affect the Magnitude of Strain in Solder Joints

When the solder joints are held under load for period of time at the extremes of thermal cycling, the solder will flow in response to this load by a mechanism referred to as *creep*, or stress relaxation. This process proceeds more quickly as a temperature increases. Therefore, its effect is more important at the upper temperature extreme than it is at the lower temperature extreme.

Most operational thermal cycles exhibit a fairly long period, typically many hours. Under these conditions tin/lead solder joints can be anticipated to almost fully stress-relax during the high-temperature half of the cycle. Therefore, test conditions that result in a nearly complete stress relaxation during the high-temperature half of the cycle should be a better predictor of operational life than shorter cycles that do not allow for this degree of stress relaxation.

One common method for understanding the amount of damage inflicted on the solder joints during a single thermal cycle is to consider the total amount of load (stress) and resulting strain (displacement) that is inflicted on the solder joint. When we plot load versus displacement for a single cycle, the resulting hysteresis loop has an area that is equal to the total amount of work performed on the joint. Figure 3.9 illustrates hysteresis loops for two different hypothetical thermal cycles for the same solder joint cycle between the same temperature extremes, but with different dwell times at high and low temperatures. It can be seen from Fig. 3.9 that the longer cycle induces greater damage into the solder joint than does the shorter cycle.

The most common practice used for a thermal cycle testing of tin/lead alloys is to utilize a sufficiently slow thermal cycle such that nearly complete stress relaxation is achieved. Total thermal

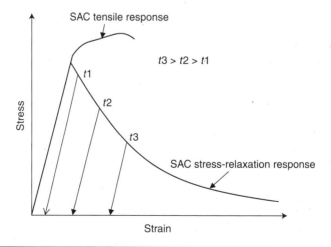

FIGURE **3.9**   Hysteresis loops for two different dwell times

cycle times of 45 min to 1 h are typically employed. This results in test durations on the order of a few thousand hours (a few months).

The most widely studied SAC alloys exhibit significantly slower stress relaxation behavior than tin/lead does under the same conditions of temperature and load. The net effect is that the thermal cycle duration that results in nearly complete stress relaxation in tin/lead will not always achieve the same results with SAC alloys.

One approach that has been considered is simply to extend the duration of the thermal cycle such that the SAC alloys will achieve a high level of stress relaxation. Unfortunately, this requires several hours for relaxation to occur at the most commonly used test temperatures. The result is that for SAC alloys, it may require a year or more to perform equivalent stress relaxation conditions, versus a few months for tin/lead. These extremely long tests are impractical to perform, and therefore alternatives approaches have been considered.

During 2006, Hillman et al. performed a comprehensive review of all the published thermal cycling test data on SAC alloys. From this review, it was concluded that a simple 50 percent derating factor could be used to account for the effects of incomplete stress relaxation during test.

More sophisticated models have been developed that account for the effects of creep and stress relaxation. One such model, the Norris-Landzderg (NL) model, shown in Eq. (3.4), was originally developed for use to predict the fatigue life of tin/lead solder joints and has been adapted for use on SAC solder joints.

$$ AF = \frac{N_O}{N_t} = \left(\frac{\Delta T_t}{\Delta T_O}\right)^{\beta} \left(\frac{t_{t}}{t_O}\right)^{\gamma} \exp\left\{k\left(\frac{1}{T_{max,O}} - \frac{1}{T_{max,t}}\right)\right\} \tag{3.4} $$

where AF = acceleration factor
N = number of cycles to failure
X_O = subscript indicates operation conditions
X_t = subscript indicates test conditions
ΔT = range of thermal cycle
t = time duration of thermal cycle
T_{max} = upper temperature limit of thermal cycle
T_{min} = lower temperature limit of thermal cycle
β, γ, k = fitting parameters (all temperatures expressed in kelvins)

Estimates of the NL exponents have been published and are listed in Table 3.5.

An example of the use of a simple power-law model and the Norris-Landzderg model is provided overleaf.

Parameter	Sn-Pb	SAC
β	2.65	1.5–2.0*
γ	0.136	0.136
k	2185	1414

*Value depends upon component type

TABLE **3.5** Published Norris-Landzderg Exponents for Tin Lead and SAC Solder Alloys

3.2.6 Calculation of Thermal Cycling Field Life Based on Thermal Cycling Test Data

Consider two very different applications where thermal cycling will be imposed on lead-free assemblies. Application 1 is a benign terrestrial application where daily cycling occurs between +25 and +60°C. Application 2 is a more stressed space-based environment with 90-min cycles between –40 and +85°C. Identical SAC 305 soldered assemblies are to be tested to determine their lifetime under these conditions. Furthermore, two different cases will be considered: one with large leadless devices and one without. Two different test temperature conditions will also be considered, 0 to +100°C and –40 to +125°C.

The test acceleration factors and the number of years of life equivalent to each 1000 cycles of test that are predicted by the power-law model and the Norris-Landzberg equation are given in Table 3.6 for application 1 (benign) and in Table 3.7 for application 2 (space).

Application 1; Benign terrestrial +25 to + 60°C daily
Acceleration factors

	SAC (w/o leadless)		SAC (w/leadless)	
	N–L	Power Law	N–L	Power Law
Test 0 to 100	22.3	17	6.3	4.8
Test –40 to +125	125.7	65.8	19.4	10.2

Application 1; Benign terrestrial +25 to + 60°C daily
Years of life per each 1000 test cycles

	SAC (w/o leadless)		SAC (w/leadless)	
	N-L	Power Law	N-L	Power Law
Test 0 to 100	61	26	17	13
Test –40 to +125	341	180	53	28

TABLE **3.6** Calculated Acceleration Factors and Life Equivalency Predicted for Application 1 (Benign Applications)

Application 2: Space-based –40 to +85 every 90 min
Acceleration factors

	SAC (w/o leadless)		SAC (w/leadless)	
	N–L	Power Law	N–L	Power Law
Test 0 to 100	0.7	0.5	0.9	0.7
Test –40 to +125	3.7	2.1	2.7	1.5

Application 2: Space-based –40 to +85 every 90 min
Years of life per each 1000 test cycles

	SAC (w/o leadless)		SAC (w/leadless)	
	N–L	Power Law	N–L	Power Law
Test 0 to 100	0.1	0.1	0.1	0.1
Test –40 to +125	0.6	0.4	0.5	0.3

TABLE 3.7   Calculated Acceleration Factors and Life Equivalency Predicted for Application 2 (Space Applications)

The result of the predictions is that 1000 cycles of either test method will be equivalent to many years of service life for the benign application, regardless of the model chosen. In contrast, 1000 cycles of either test method are equivalent to only a matter of months of service life for the highly stressed (space) application, regardless of the model chosen. These results illustrate the primary importance of understanding the actual rigors of the application environment when making service life predictions.

Other models have also been used successfully to predict thermal cycling fatigue life for tin/lead. Chief among these is the Engelmaier-Wild model, which is described fully in Appendix A of IPC-9701A. Finite element analysis techniques have also proved to be very useful in many instances for predicting tin/lead solder joint life.

3.2.7   Mechanical Shock and Vibration

Mechanical shock and vibration represent a very different environmental stress than does thermal cycling. These environments are of great importance for certain applications, but are of limited importance for other classes of applications. Table 3.8 provides a brief summary of generic application types and identifies those in which mechanical shock and vibration are important considerations.

Most of the applications where mechanical shock and vibration are of highest concern are either excluded or exempt from the European Union (EU) RoHS regulations. As a result, much less data exists regarding the shock and vibration performance of lead-free soldered assemblies than is available for thermal cycling response.

Application	Mechanical Shock is Important	Vibration is Important
Handheld devices	Yes	No
Industrial electronics	Infrequently	Yes
Automotive electronics	Yes	Yes
Avionics	Yes	Yes
Military electronics	Yes	Yes

TABLE **3.8**   Environments where Mechanical Shock and Vibration are Important Considerations

Such data as does exist almost universally indicates that SAC alloy solder joints do not perform as well as do the equivalent tin/lead solder joints in these environments, for joints to relatively stiff packages.

The reasons for this are generally understood as follows. As mentioned above, SAC alloy solder joints are significantly stiffer than their tin/lead predecessors. This can result in the generation of higher loads under the same mechanical excitation. The second underlying reason relates to the inhomogeneous microstructure exhibited by SAC alloy solders. Platelike intermetallic phases result in mechanical discontinuities that cut through the solder joints along planes. These discontinuities result in stress concentrations that can degrade the overall mechanical response of the joints.

Another critical factor affecting shock and vibration response is the nature of the intermetallic layers that form at either side of the joint. For tin/lead when soldered to copper, the intermetallic is composed exclusively of tin/copper compounds. With SAC alloys, the presence of silver complicates the composition and structure of the intermetallic layers, modifying this important interface.

Fatigue life under vibration is most commonly modeled using the (unsimplified) *Coffin-Manson model*. The median fatigue life is expressed as

$$N_f = \frac{1}{2}\left(\frac{2\varepsilon_f}{\Delta\gamma_p}\right)^{-\frac{1}{c}} \tag{3.5}$$

where N_f = median number of cycles to failure
ε_f = strain failure
γ_p = plastic strain energy
c = fitting parameter

Values for the two coefficients are well established for tin/lead. Preliminary values for SAC have also been identified as $\varepsilon_f = -0.442$

and $C = -0.57$ for SAC. The sensitivity of SAC solder joints that are formed between leadless packages (BGAs, CSPs, large chip resistors, etc.) and PCBs to early failure at high levels of vibration has important implications for reliability and testing.

Accelerated life testing for vibration typically involves increasing the vibration levels and decreasing the exposure time. Vibration test conditions that were appropriate for tin/lead are more likely to result in overtest conditions for SAC soldered assemblies exposed to high levels of vibration.

One very important environment that has resulted in much of our understanding of the shock response of SAC alloys is the dropped cell phone. It is a very common experience for users of cellular telephones to occasionally drop them directly on to pavement from a height of approximately 1 m. Manufacturers of cellular telephones discovered early during the transition to lead-free electronics that BGA devices that were attached using SAC alloy solder tended to fall off the internal circuit cards under this particular stress. Poor resistance to fracture during shock is particularly notable when SAC 305 solder spheres are used in conjunction with nickel metallization. An example of fracture that has occurred between a SAC 305 solder sphere and the nickel under-bump metallization (UBM) of the BGA is shown in Fig. 3.10. This is a scanning electron micrograph of a metallographic cross section through a SAC 305 alloy solder sphere that has fractured under shock loading at the intermetallic layer along the nickel UBM.

One proven method to greatly enhance the mechanical integrity of BGA assemblies is the use of underfill, and more recently perimeter bonding. This approach is not without its difficulties, however. There are costs associated with the dispensing of the underfill or bond material. At least as important is the fact that it is extremely difficult, if not impossible, to rework such devices. These considerations render the use of underfill and perimeter bonding impractical for many applications.

This led to extensive research into methods to address the problem of dropped cell phones without resorting to the use of underfill. The resulting solution was to greatly reduce the silver content of the SAC alloy solder spheres on the BGAs from approximately 3 percent (SAC 305) to approximately 1 percent (SAC 105). This reduction decreased the deleterious effects of silver on the bulk microstructure and on the intermetallic layers. The net result is a significant improvement in the ability of these assemblies to survive mechanical shock.

If the reduction of silver from 3 percent to 1 percent is beneficial, then perhaps complete elimination of silver would also solve the problem, maybe even better. The use of tin/copper eutectic solder has always been one of the main options under consideration for lead-free alloys. However, tin/copper suffers from two principal process-related deficiencies when compared to SAC alloys: higher reflow temperature and a greater propensity for copper dissolution. These problems have

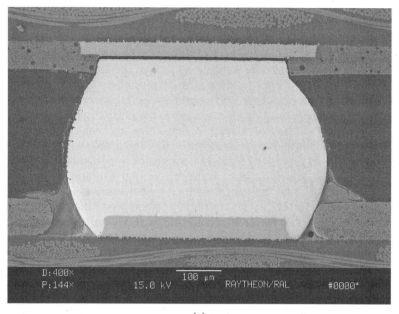

(a)

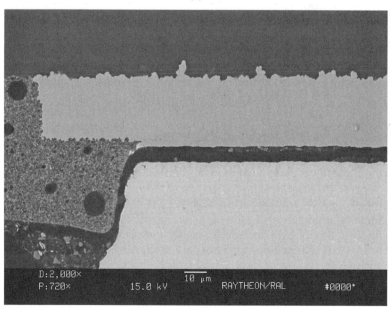

(b)

FIGURE 3.10   Failures of SAC 305 solder sphere and the nickel UBM of the BGA. (a) SAC solder sphere separated from nickel under-bump metallization as a result of mechanical stress. (b) Detail of the separation which can be seen to occur between the intermetallic in the under-bump metallization.

limited the popularity of tin/copper relative to SAC to date. A family of nickel-stabilized tin/copper solder alloys have been developed and patented by Nihon Corp. called SN100C, and they have been specifically formulated to mitigate against copper dissolution. This class of alloys appears to offer a lead-free solder option that can provide improved shock and vibration performance. Testing is currently underway to better characterize the reliability provided by this particular class of alloys. Whether the mechanical benefits provided by this alloy are sufficient to offset the problems associated with increased processing temperature remains to be demonstrated.

The issue of survivability in the face of high-vibration environments has yet to be fully addressed for lead-free solder assemblies. One reason for this is that nearly all the applications that involve high levels of vibration are either exempt or excluded from current EU RoHS regulation. Most of the industries that manufacture equipment for these environments are continuing to utilize tin/lead solder. Given the fact that such data as do exist indicate that a transition from tin/lead to SAC alloy assembling will result in a degradation of robustness in shock and vibration, it is very likely that existing electronic designs in these industries will continue to be produced using tin/lead. The reason is that the costs associated with system qualification and certifications can be prohibitively high. Therefore there is neither a regulatory nor an economic incentive for these designs to make the transition at this time.

In the future, it is likely that new designs within these industries will begin to incorporate lead-free assemblies. As the supply of tin/lead BGA and CSP packages dwindles, there will be an incentive for these high-end users to consider designing with lead-free and then performing qualification certification testing on these assemblies. However, such a transition will require many years to complete, and it is very likely that tin/lead soldered assemblies will continue to be manufactured within these exempt and excluded industries for many years to come.

3.2.8   Other Considerations: New Failure Modes

Initially, predictions of the reliability impact of the transition from tin/lead assembly to lead-free assembly were focused on those failure modes that had previously been experienced with tin/lead assemblies, such as thermal cycling and mechanical shock and vibration. It is possible that new failure modes will arise which have not been experienced or anticipated. At least two new failure modes have been reported recently that are unique to lead-free assemblies: crater cracks and champagne voids, discussed below.

3.3   The Challenge of Rework, Repair, and Reliability

Rework and repair are operations that occur after the normal manufacturing process has been completed. These actions are undertaken to reverse the effects of defects or damage to the assembly. The term *rework* is commonly applied to touch-up operations that are performed in the factory. *Repair* generally refers to operations that occur sometime after the assembly has been put in the field and has suffered from a failure or other operational anomaly.

Rework processes are more difficult to control and implement than are the standard SMT and through-hole processes used during initial assembly. Localized heating is required to reflow solder in the area where rework is performed, with a possible higher peak temperature, resulting in more severe thermal gradients and greater stresses being placed upon the board and components. These stresses create the potential for damage, causing either an immediate failure or latent failures.

A typical PCB can include thousands of individual solder joints. The reliability of the entire assembly will be determined by the reliability of the least robust solder joint. The solder joints that are formed by the rework process may not necessarily exhibit the same properties as solder joints formed by a nominal first-pass process; and in general, reworked components are less robust in the face of environmental stress than nonreworked components. Thus the overall reliability of a reworked PCB will be determined by the reliability of the reworked joints and components after rework, rather than by the reliability of the joints and components produced by the normal reflow processes.

The transition to lead-free electronics profoundly affects both rework and repair, which in turn has significant impact on reliability. Assemblies utilizing SAC alloy solder will experience significantly greater stresses during rework and repair than would similar assemblies that utilized tin/lead solder. This is directly related to the increased processing temperatures required to reflow the higher-melting-point SAC alloy solder. It is therefore essential when performing systems tests and reliability assessments to include a representative number of reworked items so that the deleterious effects of rework will be properly included in the results.

3.3.1 Rework-Induced Damage and Associated Failure Modes for Printed-Wiring Boards

Excessive localized heating of the PCB can result in discoloration on the surface and delamination of internal layers. Discoloration of the surface provides an obvious visual indication, and therefore should result in reliable detection, so will create a problem of yield rather than of reliability. Internal delamination, on the other hand, is virtually undetectable and therefore creates a risk of latent failure. Thermal stress can also result in cracking of internal vias.

Internal delamination can result in failure by two different modes: fracture of internal runs resulting in opens and the growth of conductive anodic filaments (CAFs) that results in internal shorts.

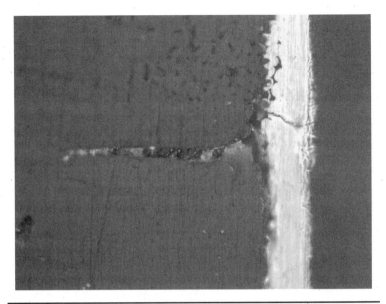

Figure 3.11 Photomicrograph of a cross section of a printed-wiring board via wall.

Delamination between internal layers results in stress concentrations when this delamination encounters a via which runs normal to the laminate. This can result in fracture of the via wall, as shown in Fig. 3.11. The stress concentration can also sharply reduce the via robustness in the face of prolonged thermal cycling, potentially resulting in intermittent open connections. Cracking of the vias can result without attendant delamination.

CAFs can grow wherever there is an open pathway between two conductors that are held at different DC potentials over a period of time. This is a particularly insidious failure mode because of its time dependence. Failures will not be detected during initial tests or even during most burn-in conditions. Failures will typically only occur after a significant duration of powered operation, which usually implies field failures.

The risk of this CAF formation increases with increasing temperature, humidity, and electric field strength. The electric field strength is proportional to the potential and inversely proportional to the conductor spacing. Assemblies that dissipate significant amounts of power, and therefore run hot, are particularly susceptible to CAF. Assemblies that operate in harsh environmental conditions that include heat and humidity are also particularly susceptible to CAF. Refer to "Cathodic Anodic Filamentation Test (CAF)" in Sec. 8.4.3 for more information about CAF testing.

Direct detection of delamination can only be performed by cross section. The ability of internal vias to survive under thermal stress can only be determined by testing assemblies using extended thermal cycling, which is both time-consuming and destructive. Susceptibility to CAF can be determined by performing extended humidity testing under DC bias, which is also destructive. Therefore, there is no practical means of inspecting reworked assemblies to verify that internal delamination has not occurred. The best means to avoid this failure mode is to properly define and control the rework processes so that the delamination does not occur. Some amount of destructive testing will be required to validate that rework processes do not induce board delamination.

PCBs can also be compromised during rework through the dissolution of copper from runs and vias into the solder. Solid copper dissolves readily into molten tin alloys. The rate of this dissolution increases rapidly with the temperature of the melt and with the amount of tin in the alloy. For this reason this phenomenon is far more important with SAC alloys than it was with tin/lead alloys. Some amount of dissolution will always occur during the initial soldering process. However, this phenomenon is far more important in the context of rework because of the increased temperature during rework, and because of the possibility of multiple rework cycles at the same location. In fact, this phenomenon is typically the limiting factor determining the number of rework cycles that can be performed at a given location using SAC.

The portion of the PCB that is most vulnerable to dissolution is the corner between runs on the surface of the board and cylindrical vias that penetrate through the thickness of the board. The dissolution rate is higher at these locations, and these features also happen to be the natural stress concentrators.

If the dissolution is so severe that immediate electrical opens are created, the problem becomes one of yield rather than reliability. However, if dissolution progresses to a point where the interconnection between the runs and the vias is nearly dissolved away, but a tiny portion remains intact, a latent failure may have been created. The most likely environmental stress that will cause separation between a via and a run is thermal cycling. During thermal cycling the PCB laminate material will expand and contract at a rate that is higher than that of the copper via. During each hot cycle the copper will be placed in tension, and fatigue damage will accumulate over time. Thinning of the copper at the stress concentration at the top of the via will drastically reduce the thermal cycling lifetime relative to nonthinned vias.

The presence of excessively thinned via connections will be nearly impossible to detect nondestructively in production electronic systems. It will therefore be necessary to use destructive means to assess the impact of rework processes on vias so the proper limits can be placed on rework for particular PCB applications.

3.3.2   Rework-Induced Damage to Components

Component-level damage can also be induced through the stresses of rework. Again, there is a significant history with tin/lead processing to identify the high-risk areas. Since the principal stresses induced into the components during rework result from thermal shock, the higher processing temperatures associated with SAC alloys create a higher risk of component-level damage.

Cracked Capacitors

Thermally induced cracking of ceramic components, particularly multilayer capacitors, has been a well-researched topic. It is not uncommon for damaged chip capacitors to function properly for an extended time, only to fail later, when the system is being operated by the customer. Once a ceramic material has been cracked, this crack can propagate under much lower loads than would be required to initiate a new crack. For this reason, chip capacitors that have suffered small cracks during rework are at risk for continued crack propagation under the effects of thermal and mechanical stress. Even when a crack propagates through the plates of a capacitor, immediate failure is not always the result. Typically, these cracks occur near one termination, so the only immediate effect is a very small change in the capacitance, which typically does not result in detectable dysfunction. However, the presence of the crack bridging between the capacitor layers compromises the interlayer insulation. Under the action of prolonged DC bias and heat, metal from the conductive layers will migrate along the crack, eventually shorting out the capacitor. Figure 3.12, which is a photomicrograph of a cross section of a multilayer ceramic capacitor that has cracked as a result of thermal stress, shows a cross section of a cracked capacitor that failed in this manner.

High-voltage capacitors operating at elevated temperature are particularly susceptible to this failure mode. Therefore PCBs that are used in power supplies or power converter applications are prime candidates for this failure mode.

It is impossible to nondestructively detect the presence of thermally induced cracks in capacitors. This creates a situation where the only means of avoiding this failure mode is to develop and validate rework processes that do not induce this type of damage. Validation will require the destructive evaluation of a statistically significant number of components.

Plastic Encapsulated Microcircuit Delamination

Many modern microcircuits are packaged using an epoxy-based molding compound. These molding compounds have been developed specifically for this application, and have been reformulated for use in lead-free solder processes so that they can withstand the higher temperatures. A failure mode was discovered early on in the implementation of this class of components that

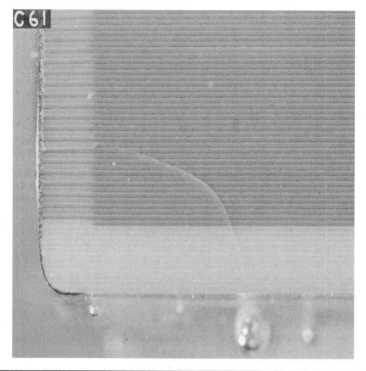

FIGURE 3.12   Cross section of a multilayer ceramic capacitor that has cracked as a result of thermal stress.

involved separation of the encapsulant from the surface of the die. This separation can lead to failure by two different modes. If the separation results in relative movement at the interface, the bond wires that are attached to the top of the die can be torn asunder, resulting in an immediate failure. If the separation is small, the bond wires may remain intact. However, the resulting crevice can leave the device susceptible to later attack by moisture intrusion. The presence of moisture, DC bias, and any ionic species can lead to corrosion at the bond pad bond to wire interface, or to metal migration shorting on the surface of the die.

Delamination of plastic encapsulated microcircuits during solder processing can result from rapid evolution of the water that is dissolved into the epoxy. Catastrophic delamination by this means is referred to as *popcorning*. Sensitivity of packages to this failure mode has been studied extensively and is covered by JEDEC Standards 20 and 33, which defines standard *moisture sensitivity levels* (MSLs) for use in describing these packages and their appropriate handling protocols. The applicable MSL is a function of the peak temperature to which the device is exposed. Therefore, a device that will withstand a tin/lead reflow process at 205°C without popcorning may not survive a SAC reflow process at 240°C. For this reason, component manufacturers now rate their components for moisture sensitivity of the higher reflow temperatures experienced during SAC processing. Since components are likely to see higher processing temperatures during rework than during normal soldering, this is the manufacturing step where there is the greatest risk of popcorning.

In addition, each MSL rating defines a period of time that components may be exposed to moisture prior to soldering, before rebake is required. During a well-controlled manufacturing process the time when such parts are removed from dry moisture-sealed packaging is recorded. Thereafter, the parts are tracked to ensure that the MSL requirements are satisfied. Rework and repair conditions create a new set of challenges with regard to this issue. Assemblies requiring rework and repair have typically been exposed to uncontrolled moisture levels for extended periods. Moisture-sensitive parts in the vicinity of a rework site may be at risk for popcorning. Bake-out of assemblies to be reworked or repaired is recommended, unless the MSL requirements of all potentially affected components are understood together with the relevant history of the assembly's exposure to moisture.

Catastrophic delamination that results in prompt electrical failure can readily be detected during in-circuit test or other electrical functional testing. Delamination that does not result in prompt electrical failure will be quite difficult to detect. The most common technique used to detect and measure delamination in plastic encapsulated microcircuits is *scanning acoustic microscopy* (SAM), which is a form of ultrasonic imaging.

This technique requires that the unit under evaluation be fully immersed in a working fluid, typically water (although isopropyl alcohol is also used occasionally). Therefore, this technique is typically not employed to detect delamination of microcircuits that have already been soldered onto deliverable assemblies. This technique is best used to validate processes that have been developed, to ensure that delamination does not occur during rework operations.

Effects of Rework and Repair on the PCB Assembly

The rework and repair process involves localized heating of the device which is typically removed and replaced. This inevitably results in subjecting all the adjacent components to some amount of heating and potentially severe thermal gradients. Quite often it is these components that suffer damage during the process, and not the device that has been replaced.

The large thermal gradients that are present in the region surrounding the rework can lead to melting of a portion of the solder joints of a large device. This can create latent stresses in the assembly that will eventually result in failure. The reasons for this are as follows:

During normal solder attach processing, all the solder joints on a given device will solidify in approximately the same time. During this time frame, there should be only small thermal gradients across the device and across the region of the PCB beneath the device. The situation results in self-centering of the component relative to the solder pads. Upon cooling there will be low levels of built-in stress, which is then distributed across the joints in a well-behaved manner.

If only a portion of the solder joints of a device are remelted during a rework operation on an adjacent device, there will be significant thermal gradients across both the component and the board. The solder joints will solidify one by one in a sequence from the cool side toward the warm side. The thermal gradients will result in complex three-dimensional differential movement between the board and the component, with each joint frozen in the condition imposed upon it at the time of solidification. The result will be increased stresses built into the joints, and suboptimal alignment between the pads and the device. The solder joints will be substantially less robust in the face of mechanical stress induced by thermal cycling or shock and vibration than they were prior to the rework operation.

It is therefore vital that rework processes be evaluated for their effect not only on the target component, but also upon all adjacent components. The overall reliability of the assembly will very likely be determined by the condition of the assembly in this "heat-affected zone."

3.3.3   Special Considerations for Repair

In most commercial applications, failed PCBs are simply discarded and replaced. Therefore, issues of solder and soldering are of no concern relative to repair of such systems.

Repair of electronics systems down to the level of component replacement on PCBs is confined to a limited range of industries. Only military, aerospace, and medical electronics are typically repaired to this level. These systems are often maintained for periods of 10 or 20 years, and repair activities may occur at locations quite remote from the initial manufacturing facility. Repair of this sort has always involved special challenges. The advent of lead-free electronics will complicate these challenges.

Repair is nearly identical to rework except that it is typically performed after the system operated for some time in the field. All discussions on lead-free rework are similarly applicable to repair operations.

Repair operations introduce a unique set of additional concerns. The first concern relates to determination of the solder alloy that was originally used to manufacture the assembly to be repaired. In the past, all electronic assembly was performed using essentially identical tin/lead alloys. All assemblies could therefore be repaired using the same generic processes. The transition from tin/lead to lead-free assemblies has created a much more complex world for the repair depot.

Repair depots may be receiving a mix of lead-free and tin/lead assemblies. Furthermore, the lead-free assemblies may have been assembled using a range of lead-free alloys. The use of repair procedures developed for use on one alloy system may not be generally applicable for another alloy system. The best possible outcome of the current investigations is that a wide range of assemblies will be repairable by a single repair alloy, or by a very small set of repair alloys. This will greatly decrease the logistical burden placed upon remote repair depots.

However, the scenario creates a complex situation from the viewpoint of those who must ensure qualification and certification of repaired equipment. The traditional approach to repair is that it returns the equipment to its original intended configuration, as specified in the applicable technical data package. This configuration will have been qualified and/or certified by customers, and regulators are responsible for ensuring the safety of the public. If the solder alloy used during repair does not match the solder alloy used during initial manufacture, some regulators might consider that the qualification/certification of the equipment has been invalidated. This potential conflict between the practical needs of the repair community and the requirements of the regulatory community will have to be resolved in the future.

3.4   Lead-Free Solderable Finishes

Tin/lead alloys were the most common solderable finishes in use both for component terminations and for PCB pads prior to the conversion to lead-free electronics. With the mandated elimination of lead, these finishes have been replaced by a variety of lead-free alternatives. Some of the common lead-free alternatives are listed in Table 3.9, together with notes on potential reliability impacts.

3.4.1   Pure Tin Finishes and Tin Whiskers

Unfortunately, pure tin finishes are susceptible to an unusual phenomenon referred to as *tin whiskers*. Whiskers are elongated single-crystal filamentary growths of pure tin that have been reported to grow to more than 1 cm (250 mils) in length (although they are more typically 1 mm or less) and from 0.3 to 10 µm in diameter (typically 1 to 3 µm). Whiskers require no special environmental conditions to form. They can grow spontaneously without an applied electric field or moisture (unlike metal-migration dendrites) and without atmospheric pressure (they have been known to grow in a vacuum). Whiskers exhibit many shapes — straight, kinked, hooked, or forked — and some are even reported to be hollow. Their outer surfaces typically exhibit striations running along their lengths. Whiskers can grow in nonfilament types which are sometimes called *odd-shaped eruptions* or OSEs. Some examples of tin whisker growths are shown in the electron micrographs in Fig. 3.13.

Finish	Applications	Potential Reliability Issues
Electrodeposited tin	Component terminations	May grow whiskers
Immersion tin	PWB pads	Can grow whiskers?
Immersion silver	PWB pads	Champagne voids? Electromigration?
Electroless nickel immersion gold (ENiG)	PWB pads	Black pad
Electroless nickel immersion palladium immersion gold (ENiPiG)	PWB pads	Limited field experience
Hot dip SAC	Component terminations	Can grow whiskers?
Electrodeposited tinbismuth	Component terminations	Can grow whiskers?
Hot air solder level (HASL) SN100C	PWB pads	Limited field experience
Nickel/palladium/gold (TI process)	Component terminations	None

TABLE **3.9** Common Lead-free Solderable Finishes

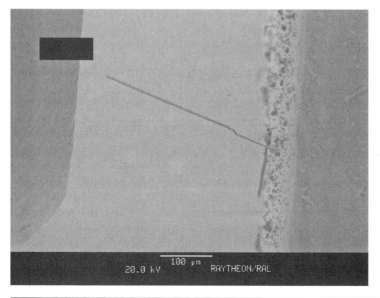

FIGURE **3.13** Examples of tin whisker growth imaged using electron microscope.

Whisker growth kinetics are generally described as occurring in three phases: incubation (or dormancy), growth, and saturation (or termination). Whiskers typically do not form immediately after the plating process. The period of time between plating and the initial formation of whiskers is the incubation or dormancy phase. This phase has been reported to be as short as only a few minutes to as long as several years. The most common incubation periods range from several months to a few years in length.

Whisker growth rates also vary over a very wide range. Growth rates between 0.03 and 9 mm/yr have been reported. Following a period of growth, most tin whiskers will actually stop growing. However, it is not known whether these whiskers can start growing again if there are changes to the environment that might stimulate regrowth. Also, there are some studies that have examined the growth of whiskers after many years of growth that have revealed continued growth. Therefore, there is no age beyond which a tin-plated surface may be considered to be immune from further whisker growth potential.

Tin whisker growth mechanisms have been studied for several decades. As yet, there is no single explanation of this mechanism that has been widely accepted. Tin whisker growth is most often attributed to stresses in the tin deposit. These stresses can arise from a wide variety of sources. These sources include residual stresses in the tin resulting from the plating process, formation of intermetallic compounds (most commonly copper/tin intermetallic compounds), especially within the tin grain boundaries, and coefficient of thermal expansion mismatch between the plating material and substrate. Another important source of stress is externally applied compressive stresses, such as those introduced by compression on the surface by a nut or a screw, bending or stretching of the surface after plating, and scratches or nicks in the plating introduced by handling. Many reported tin whisker failures have been associated with regions of the plating where fasteners or contacts have impinged upon the surface of the tin.

Tin Whisker-Induced Failures

Tin whiskers have been reported to cause failures in electronics by creating short-circuit pads between adjacent conductors. A wide range of products and applications have been affected including satellites, nuclear power plants, missile systems, heart pacemakers, and wristwatches. The kinetics of whisker growth result in a potential latent product failure mode. While products are being manufactured and tested, the whiskers are typically still incubating. Growth then commences after the product has been delivered to the customer, and most tin whisker-induced failures are reported to occur after 1 to 3 years of service.

The propensity for particular tin deposits to grow whiskers of sufficient length and quantity to induce failures varies dramatically.

Experience indicates that the tin whisker propensity is strongly dependent on plating process conditions. Therefore, whiskers will tend to affect specific lots of material disproportionately. The net result is that tin whisker problems are not experienced at a low level uniformly distributed across all tin-plated components. Rather, problems erupt in localized "rashes" where a cluster of failures is associated with a particular component in a particular application.

For this reason, tin whisker-induced failures have an effect that is out of proportion to their overall average rate of occurrence. Failures that do arise typically occur when systems are in the field, during a portion of the system lifetime where failure rates are anticipated to be at a minimum. Since the failures will affect individual lots of components, which have been distributed among each of the products in the field, the failures will tend to affect multiple fielded systems. The net effect is that rather than the occasional failed component, tin whisker-induced failures create the occasional failed product line or program.

Under normal atmospheric conditions, tin whiskers exhibit a limited current-carrying capacity, typically in the range of 5 to 32 mA, although currents as high as 75 mA have been reported. Because of this limited capacity, tin whiskers will fuse open under a wide range of typical voltage conditions in a very short time (microseconds). For many applications, this implies that tin whisker shorting will have no immediate effect on functionality. A good example of this is an instantaneous short between voltage supply input and ground. A short-duration voltage glitch will appear on both ground and the voltage supply line. However, most properly designed systems include filtering capacity that will damp out most such transients, eliminating any immediate impact. In other situations, these short-circuits can have serious system-level consequences: an example is a short between a voltage line and a reset line that caused a transient pulse which was interpreted by the system as a reset signal, and the entire system shut down. Since the conductor fused open during the failure event, it was extremely difficult to identify the source of this recurring intermittent problem. The potential exists for transient tin whisker-induced shorts to produce a wide range of apparently random, intermittent faults in digital systems.

It is therefore widely believed that the majority of tin whisker-induced failures are transient in nature, and therefore not properly diagnosed, due to lack of experience by technicians and engineers with observing whiskers. Whiskers can be extremely difficult to detect visually. Even with a good stereo microscope, whiskers can be virtually invisible if the lighting conditions are not perfect. In addition, if an operator has never seen a whisker, the presence of filaments on the surface may be ascribed to the presence of dust, or other foreign material of no consequence.

Tin whisker-induced short-circuits can be persistent under certain circumstances. If the available current is limited to less than the fusing current of the whisker, a persistent short-circuit will evolve. Plasma can form if there are significantly high levels of voltage and current available. This plasma may persist and can conduct large amounts of current on the order of tens of amperes. The power dissipated by this event will consume metal from either side of the short, continuing to feed the plasma. The initiation and persistence of this plasma are highly dependent on the pressure of the gas present. Initiation and sustainment of tin plasma have been demonstrated using a power supply voltage of only 4 V in vacuum; and with 28 V under conditions of 1 atm of both air and nitrogen. These voltages may be unusually low when compared with theoretical numbers for initiation of plasma arcs, and can be explained by short-duration voltage spikes occurring during the initial shorting event, producing sufficient excitation to initiate the plasma arc. The use of tin plating is of particular concern in space-based vacuum applications.

Tin whiskers also pose a potential reliability problem if they break away from their point of origin and migrate to other locations where they may cause short-circuits. While such occurrences may be rare, such failures would be extremely difficult to diagnose. The offending whisker would likely have fused during the failure event or have become dislodged during the analysis process.

Short-circuit failures induced by migration of liberated whiskers have been widely reported in zinc electroplating, which is also susceptible to whisker growth. Zinc was widely used as a surface finish on steel floor panels installed in raised floors of data processing centers. The most common reported scenario involved a maintenance activity when the floor panels were removed and replaced, causing a large number of zinc whiskers to break off. Airflow within the data center then transported these whiskers into the computer systems, resulting in random intermittent failures. Mitigation of these conditions has spawned a specialty cleaning industry of removing floor tiles and vacuuming away the zinc whiskers.

This experience with zinc whiskers can be instructive for assessing the risk posed by migration of tin whiskers. The surface area of plated zinc on the floor tiles was on the order of several square meters, vastly greater than the surface area of tin plating in any reasonably sized electronic system. Liberation of the zinc whiskers required direct physical contact with the surface of the zinc to break the whiskers off. Tin plating used on most solderable finishes of electronic components and PCBs should not experience direct mechanical contact during any normal portion of their operational life. However, pure tin plating is sometimes specified for use for other applications including mating surfaces for electromagnetic interference (EMI) shields. Such applications could result in significant areas of pure tin that are subject to periodic mechanical impingement and could liberate large quantities of whiskers.

Under most circumstances, applications of tin plating where there is no metal surface within a centimeter of the tin can be considered as immune from tin whisker-induced failures. However, if large areas are plated with tin and these areas are subject to mechanical impingement, electronic assemblies at a distance may be at risk for failure.

Given this history, many have questioned the wisdom of the component industry's decision to implement pure tin finishes on a broad basis. Part of the rationale for this decision was optimism that in the years leading up to the implementation of RoHS that an effective mitigation strategy would be identified and implemented. Currently, the only proven mitigation technique is to alloy the tin with a minimum of 3 percent by weight of lead. Unfortunately, the RoHS limit on lead is 0.1 percent by weight of lead. The interpretation of the regulation that this applies to the plating content separate from the component means that this only proven mitigation strategy is barred in many cases.

An exemption has been granted by the EU to permit the use of lead as an alloy constituent in tin for the purposes of tin whisker risk mitigation when used on a device with very small lead pitch. This exemption was only granted a few months before the July 2006 deadline. As a result most component manufacturers have already converted their product from tin/lead to pure tin. The net result is that the only components that are available from component manufacturers in many instances incorporate pure tin terminations.

Tin Whisker Risk Mitigation

A large fraction of component vendors supply their components only with pure tin solderable terminations. Therefore, most OEMs are compelled to incorporate pure tin into their systems. It is therefore important to employ such mitigation techniques as are available to reduce the risk of whisker-induced failures.

The only proven mitigation technique is to alloy the tin with a minimum of 3 percent lead by weight. All other mitigation techniques may reduce the risk of whisker-induced failures, but cannot be considered to totally eliminate the risk. The most commonly employed tin whisker mitigation strategies are summarized in Table 3.10. The efficacy of each of these strategies is discussed below.

Lead Alloy Content While it is widely accepted that 3 percent by weight of lead alloyed into the tin is an acceptable mitigation for all applications, including for use in the highest reliability systems, there are reports that the mitigating effects of lead are fully developed at concentrations as low as 1 percent by weight. Below 1 percent by weight the mitigation effects drop off, at a rate that is not well documented. Unfortunately, there are no data indicating that levels below 0.1 percent by weight (the RoHS-mandated limit) provide any meaningful level of mitigation.

Strategy	Application	Limitations
Avoid use of pure tin finished components	Design, component selection	Increasingly difficult as component suppliers transitioned to pure tin
Cover pure tin with tin/lead	Component reprocessing, circuit card assembly	Adds cost and may not be possible under all circumstances
Increase gap size between tin and adjacent conductors	Design, component selection	Limited by available component geometry and circuit card area constraints
Use nickel underplating	Design, component selection	Many components not available with this feature
Use postplate-bake-out	Component selection	Not all component suppliers perform this
Reflow plating	Component selection, circuit card assembly	Not always achievable during circuit card assembly, rarely available from component suppliers
Use conformal coating	Circuit card assembly	Increased cost, complicates rework
Use potting compounds	Higher-level assembly	Increased cost, complicates rework

TABLE **3.10**   Common Tin Whisker Mitigation Strategies

For applications where compliance with our RoHS limits is not required, lead is sometimes intentionally added on to pure tin finishes to provide tin whisker risk mitigation.

One common means of replacing the pure tin with a tin/lead alloy is to dip the tin surface into molten tin/lead. This can be performed at the component level in many cases through a process referred to as *robotic solder dip*. Commercial component reprocessing service providers offer robotic solder dip on a routine basis. Manual solder dip has also been performed, but most high-reliability users (who are the ones most likely to resort to such nonstandard, costly methods) typically do not trust manual solder dip processing to provide damage-free components.

A proprietary process for use on chip-style components has been developed, and is offered by AEM, Inc. In this process, lead is electroplated onto the terminations directly over the tin. In a second step, the components are heated to promote interdiffusion to yield a relatively homogeneous coating of tin/lead.

PCBs using tin/lead solder will always result in some coverage of any pure tin solderable component terminations by tin/lead, mitigating whisker growth in these regions. In some instances, the tin/lead solder will cover 100 percent of the original tin surface, thereby fully mitigating the tin whisker risk. However, whether this perfect coverage is actually achieved will depend upon the geometry of the termination, the geometry of the PCB metallization, and the soldering process parameters. Incomplete coverage of the component termination will leave a portion of the lead susceptible to whisker growth. A very infamous tin whisker-induced failure occurred under exactly such conditions. The tin-plated through-hole leads of a device were soldered using tin/lead, which did not travel all the way up to the body of the device. Tin whiskers grew between the leads, shorting them out. The net result was that a nuclear power plant in New York State was off-line for 6 months. This failure illustrates the importance of careful evaluations of the ability of tin/lead solder to completely cover tin for specific applications, if this approach is to be relied upon.

Increased Conductor Spacing   Conductor spacing affects the risk of tin whisker shorts because only whiskers longer than the conductor spacing can physically bridge the gap. Any collection of actual whiskers will contain whiskers of various lengths between zero and some maximum value. Therefore, the risk of tin whisker shorts will always decrease as conductor spacing increases.

The distribution of whisker length and orientations has been studied in a limited number of instances, since this work is extremely painstaking. All the studies to date indicate that short whiskers are far more numerous than longer whiskers. One study suggests a lognormal distribution of whisker lengths.

If the risk of a short across a given gap is assumed to be proportional to the number of whiskers of sufficient length, then the risk as a function of gap size can be derived directly from the probability distribution of whisker lengths. The risk of failure for a specific gap size will be proportional to the integral of the probability density from that length to infinity. In statistical parlance, this function is referred to (ironically in this case) as the *survival function* $S(x)$. For a probability distribution of whisker lengths that is described by a lognormal function, the survival function will drop off very rapidly with increasing length/gap. The net result is that one of the most important factors in mitigating whisker risks is the conductor spacing.

Unfortunately, the minimum conductor spacing is often determined by the geometry of leads on a particular component, over which designers and manufacturers have no control. Therefore, conductor spacing is usually more useful to consider when performing risk assessments, than when developing mitigation strategies.

However, if designers have the choice between package styles with differing interconductor spacing, choosing the style with the larger interconductor spacing will significantly mitigate the risk of tin whisker failures.

Conformal Coating   The use of organic conformal coating compounds is fairly widespread in the military and aerospace electronics industries, but is far less common with commercial and consumer electronics. These coatings have been developed to protect PCBs from environmental degradation, particularly humidity and corrosion. As a secondary benefit, these coatings also provide some protection against shorting induced by foreign objects.

The use of conformal coating is a very attractive option for high-reliability OEMs. It is an existing process that is under their control and can be applied atop a wide range of components and assemblies. However, conformal coating materials were not developed, and are not typically tested, for their ability to provide tin whisker risk mitigation. Recent experiments have been performed, and models have been developed, to ascertain the amount of mitigation that these coatings provide.

A study was performed by NASA indicated that some level of tin whisker risk mitigation was provided by the use of a particular urethane conformal coat. In addition, an analysis was performed that indicated that tin whiskers that pass through open space between two conductors should not be capable of penetrating conformal coat on the distant conductor, but instead would buckle prior to penetration. More recent studies have also focused on improved modeling of the whisker buckling phenomenon.

This tin whisker buckling phenomenon has been observed in the laboratory. The whisker shown in Fig. 3.14 was seen to have buckled after contacting a very soft silicone potting material. Penetration of the whisker tip into the silicone was so infinitesimal that the buried tip of the whisker was still observable using the scanning electron microscope. This image shows a long tin whisker which buckled when it continued to grow after contacting silicone potting material at its tip.

Meaningful assessments of the degree of mitigation provided must take into account the performance of actual coating systems and actual circuit geometries. Assumptions of perfect coating performance will result in overly optimistic risk assessments. On the other hand, excessive concern regarding whisker growth mechanisms that cannot result in electrical shorts will result in overly pessimistic risk assessments. It is important to be mindful of the fact that conformal coating, like other strategies, only provides mitigation against whisker-induced failures, and not 100 percent prevention. What needs to be assessed is how much mitigation a particular real coating system will provide in actual circuit applications.

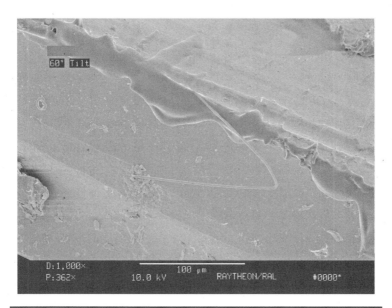

Figure 3.14 Tin whisker buckling.

A conformal coating that is applied onto real circuits will result in variable coating thickness, with some isolated areas exhibiting zero thickness (void). In addition, while data indicate that whiskers can penetrate conformal coating, they also show that the number of whiskers that penetrate a coated surface is only a fraction of the number of whiskers that grow on a similar uncoated surface.

A model has been published by the author that quantifies the amount of tin whisker risk mitigation provided by conformal coating, based upon properties of the coating. This model predicts that most conformal coating should provide at least a 100:1 reduction in the incidence of whisker-induced shorting. One key assumption in this model is that the overall whisker density is not so large that the probability of two whiskers growing from opposite leads and meeting in the middle is nil. Recent work was performed using Monte Carlo techniques to evaluate the likelihood of such whisker-to-whisker shorts. It was concluded that under conditions of high whisker density the probability of such events may not be negligible.

Nickel Underplating A nickel plate directly below the tin has been used for many years as a method of tin whisker risk mitigation. It has been shown that the whiskers that grow from tin that has nickel underplating tend to be much shorter than are whiskers that grow from tin deposited directly atop copper substrates.

Unfortunately, the addition of nickel underplating is not a drop-in replacement for tin plating directly over copper. The presence of nickel underplating affects lead forming and trimming operations, and therefore affects a considerable number of package manufacturing steps. As a result, it is not typically possible for an end user of components to impose a requirement of a nickel underplating on a component vendor. If alternate components are available that incorporate a nickel underplating, their use will greatly reduce the risk of tin whisker-induced failures.

The mechanism by which nickel underplating reduces the growth of long tin whiskers is currently understood as follows: The nickel serves as a barrier to diffusion of copper to the tin, and therefore eliminates the growth of copper/tin intermetallic compounds. Copper-tin intermetallic compounds have been associated with the formation of tin whiskers. Nickel reacts with tin to form intermetallic compounds very differently than does copper with tin. The rate of reaction is much slower, so the resultant accumulation of intermetallic compounds over the same time and under the same temperature conditions will be much less for nickel and tin, than with copper and tin.

Significantly, when copper reacts with tin to form an intermetallic compound, there is a net increase in the volume. Therefore, when this material forms, especially along grain boundaries, compressive mechanical stresses are induced into the tin. However, when tin reacts with nickel, there is a net decrease in volume. This will result in tensile stresses, which tend to reduce the propensity for whisker formation.

Postplate Heat Treat (Nonfusing)   Thermal treatment of the tin plating to temperatures below the melting point of tin, which is performed immediately after deposition, has been shown to reduce the propensity of whisker formation in many cases. This phenomenon was first described during the 1960s, and its effect was believed to be the result of driving off dissolved hydrogen from the tin deposit. This technique was set aside in favor of the addition of lead to tin, which was discovered in the same time frame and which has shown to be fully effective as mitigating whisker growth.

Interest in this technique was revived due to the impending ban on the use of lead during the late 1990s. During this time frame the mitigating effect was usually ascribed to stress relief within the tin plating. For this reason these postplate heat treatments are now commonly referred to as *annealing*. Because these treatments must be performed immediately after deposition of the tin, they are only available to component manufacturers and cannot be employed by downstream users.

Fusing or Remelting of Tin   Complete remelting (fusing) of the tin deposits has been employed for several decades as a mitigation strategy for whiskers. The traditional approach involves dipping of the tin-plated item into a bath of hot oil. The availability of this particular process has dwindled over the years as a result of safety issues. Therefore, this approach may be of limited practical value in today's manufacturing environment. Complete remelting of tin plating cannot be presumed to occur during standard reflow assembly processes, even those designed for lead-free SAC solders.

The benefits of remelting are due to the elimination of stresses generated during tin plating. In addition, dissolved organics and other gases that contribute to whiskering propensity will evaporate during remelt. The resulting grain structure will tend to be coarser than the initial grain structure. A relatively thick intermetallic layer will form between the tin and the substrate that reduces the rate of diffusion and subsequent intermetallic formation.

Tin Layer Thickness   The thickness of the tin layer has been shown to affect whiskering propensity under certain circumstances. This effect is particularly pronounced when the tin is deposited directly over a copper alloy. For tin over copper, thicker layers on the order of 10 μm thick or more generally exhibit a lower propensity for growing whiskers than do thinner layers. This effect seems to be dependent upon the composition of the material on to which the tin is deposited. It has been reported that when tin is deposited over nickel, thinner layers exhibit a lower whisker propensity than do thicker layers.

This difference in the dependence of whiskering propensity on the composition of the substrate is likely related to the different intermetallic reactions that occur, and how they affect stress within the tin deposits. The growth rate of intermetallics is limited by diffusion transport within the solid materials. As a result the effects of intermetallic growth are most concentrated in regions close to the interface between the tin and the substrate. If the tin deposit is quite thin, then the region influenced by intermetallic growth effects could extend all the way up to the surface. On the other hand, if the tin deposit is relatively thick, the effects of the intermetallic growth may not extend all the way to the surface.

Since the growth of copper/tin intermetallic compounds has been shown to increase the compressive stress, and thereby promote whisker growth, a relatively thick tin plating tends to ameliorate this effect. On the other hand, the growth of copper/nickel intermetallic compounds has been shown to increase tensile stresses and is believed to reduce whiskering. In this case, the beneficial effects on the intermetallic growth would be more pronounced in a thinner layer.

Tin Deposition Process: "Bright" Tin and "Matte" Tin   Whisker propensity of tin deposits is related to the type of deposition process and to the details of the process employed. The overwhelming majority of solderable tin finishes are deposited electrolytically from an aqueous plating solution. Other types of deposition processes that are not commonly used on electronic components include hot tin dip, vacuum deposition, and flame spray.

Unfortunately, electrodeposited tin seems to exhibit the highest propensity for whisker growth compared to tin deposited from a hot dip. Hot dip tin may be thought of as similar to tin that has been remelted or fused. A fairly large grain structure and very low initial stress should result. There are very few data on the whiskering propensity of tin deposited from vacuum deposition or from flame spray.

Processes for electrodeposition of tin are quite varied. The detailed chemistries and processes associated with plating are closely guarded trade secrets, so it is often difficult for those researching tin whiskering to obtain all the information that may be of import for understanding how process details relate to whiskering propensity. The most basic top-level distinction among electrodeposited tin is between bright tin and matte tin.

The terms *bright* and *matte* are used quite commonly by producers of tin deposits and by users of tin deposits. However, there are a wide range of interpretations as to what these terms actually mean. The most basic distinction is in the visual appearance: shiny or dull. For some producers of metal finishes this is the only distinction that has been made.

The most common interpretation of the meaning of bright tin is that it describes tin plating that incorporates organic grain refiners. The resulting grain structure exhibits a grain size that is smaller than the wavelength of visible light. The resulting structure includes multiple grains across the thickness of the deposit. As a result the deposit appears bright and shiny and is therefore popular for cosmetic applications. The resulting deposit is also significantly harder than other types of tin deposits. This factor can be of practical importance when resistance to wear is important, such as with connector applications.

Conversely, the most common interpretation of the meaning of *matte* is a tin deposit that contains no grain refining additives and therefore has very low carbon content. The typical grain size is much larger than with bright tin. It is not unusual for matte tin grains to extend through the entire thickness of the deposit to create a generally columnar microstructure.

Generally, bright tin deposits were considered to always exhibit a higher whiskering propensity than do matte tin deposits. This is untrue, since it was shown that matte tin deposits can grow whiskers quite well, and that under some conditions bright tin will exhibit less whiskering than matte tin. In particular, when stresses are induced into a plated deposit as a result of direct pressure on the surface, the softer matte tin deposits may grow more whiskers than bright tin deposits. As always, there is extreme variation in whisker propensity between tin deposits that are seemingly identical. This variability can overcome any difference between bright and matte tin whisker properties.

Tin Whisker Test Methodologies

Since the whisker propensity of a particular tin coating can range from almost none to extreme levels of whiskering, it would be very useful if there were some means to test a particular coating to determine its whisker propensity in its intended application. Ideally, such testing would be relatively quick and inexpensive to perform. Suppliers and industry consortia teams were formed to develop models for tin whisker growth that could be used to support accelerated testing methods.

As the RoHS deadline approached, understanding and modeling of the whiskering phenomenon were not sufficiently developed to support definition of accelerated testing methods. Despite this, a set of standard whiskering tests was established, and has been codified under industry standard JESD 201. Whisker propensity testing in accordance with JESD 201 is neither quick nor inexpensive. A summary of these test regimes is provided in Table 3.11.

Approaches to Tin Whisker Risk Management

The JEDEC standard JESD 201 defines three classes of applications, with three corresponding approaches. For classes 1 and 2 there are different acceptance levels for the whiskering propensity tests that can be applied. For level 3, critical high-reliability electronics, the use of tin is not recommended. This creates a conundrum for the high-reliability industry, because most of the component suppliers offer only tin-plated terminations.

Manufacturers supplying to non-high-reliability industries may choose to utilize testing for JESD 201 to mitigate tin whisker risks. Another widely employed technique is to develop a set of design rules based upon tin whisker risk assessments that define applications where tin may and may not be used. A more complete discussion of this approach is provided in the following section on risk assessments.

Several years before the RoHS deadline, the high-reliability industry realized that the issue of tin risk management had to be addressed head on. It was clear that neither the component manufacturers nor OEMs or other industry segments were going to

Environment	Conditions	Duration
Temperature cycling	−55°C or −40 to +85°C	1000 cycles
Ambient temperature and humidity	30°C and 60% RH	3000 h
High-temperature and humidity storage	60°C and 87% RH	3000 h

TABLE **3.11** JESD Tin Whisker Test Environments

provide any direction or assistance. A tin whisker risk control working group was established within the Lead-Free Electronics in Aerospace Project-Working Group (LEAP-WG) organization. (See the details on this organization and the documents they have published included in the section on aerospace and high-reliability industry standards below.) This group was tasked to define standard system-level tin whisker risk management techniques.

One of the challenges faced by this group was to define the wide range of applications that fall under the umbrella of aerospace and high reliability. At one extreme there are space-based systems that are susceptible to the vacuum plasma arc phenomenon, and where the consequence of a single failure can result in the loss of multimillions of dollars. At the other extreme, is ground-based communications and surveillance equipment, which is regularly tested and repaired. The level of vigilance and control, and costs per unit that are appropriate for satellites, will not be appropriate for handheld radios.

An approach was devised to deal with the wide range of system realities. Several standards to control levels were defined, based on the degree of customer involvement, level of inspection, recordkeeping, and risk assessments. Higher levels require greater vigilance than do lower levels, and costs increase with increasing level. The tin control levels and their associated controls are summarized in Table 3.12. The full text requirements are incorporated in GEIA-STD-0005-2.

Tin Whisker Risk Assessment Methodologies

Designers of high-reliability electronics systems are faced with the decision of how and when to incorporate components with pure tin finishes into their products. Part of this decision-making process is to assess the risk posed by the various possible applications of tin, and then to weigh these against the costs associated with implementing alternatives.

Tin whisker risk assessments were also performed after the fact when tin had been inadvertently incorporated into high-reliability systems, where nontin finishes were specified. While precise calculations of the probability of failure as a function of time for any particular application of tin cannot be performed, yes/no decisions must be made routinely as to the suitability of tin. As a result, these decisions are made on the basis of "engineering judgment," which is synonymous with consensus of opinion.

When performing tin whisker risk assessments, it is necessary to consider the actual usage of application for the component in question. Typically, it is not possible to assess tin whisker risk by looking exclusively at individual components; rather it is necessary to consider the conditions that the components experience when they are incorporated into the system. For example, a component may be mounted in such a manner that there is a nearby component that is

	Documentation of Tin Use	Detection and Control	Mitigation	Risk Analysis
Level 1	Supplier: General information on finishes used	None	None	None
Level 2A	Supplier: General information on finishes used Customer: List of any applications where tin is not allowed	None	None. However, if mitigation methods were assumed for the purposes of analyses, the supplier shall report those assumed mitigations	At the process level: analyses showing application tolerance to whiskers, or analyses of tests to demonstrate propensity of whiskering, or field data analysis demonstrating requirements will be met even if no mitigations are applied
Level 2B	Supplier: List of families of tin-finished piece parts and categories of applications where they would like to use tin Customer: List of any applications where tin is not allowed	It is recommended that the supplier and customer develop a sampling plan for confirming materials received	It is recommended that at least two mitigation methods be employed	At the family level: analyses showing application tolerance to whiskers, or analyses of tests to demonstrate propensity of whiskering, or field data analysis demonstrating requirements will be met. Some individual uses may need to be analyzed at the instance level

TABLE 3.12   Summary Of GEIA-STD-0005-1 Tin Control Level Requirements

very close to the tin termination, which provides the highest-risk pathway for a short. Also conformal coating material will be applied at higher levels of assembly, and this will significantly affect risk assessments. Another important consideration is whether the tin surface in question will be replaced by eutectic tin/lead during the next higher assembly solder attachment.

Given the large number of tin risk assessments that need to be performed, standard methodologies need to be employed if these assessments are to be performed effectively and efficiently. There are three basic types: rules-based approaches, risk assessment algorithms, and Monte Carlo simulations. These methodologies are not mutually exclusive, and a standard process can be defined that uses a mix of these.

The basic approach for all these methodologies is to consider each of the possible mitigating effects of the factors that were described in the previous section on tin whisker risk mitigation. All pertinent factors must be considered simultaneously, in the context of the application in question. A hypothetical example of a set of design rules for use in mitigating tin whisker risks might be as follows:

1. When the minimum interconductor spacing exceeds 2 mm, the use of tin finish is always permitted.

2. When the minimum interconductor spacing is between 1 and 2 mm inclusive, tin may be used under any one of the following circumstances:
   a. The component lot has passed JESD 201 level 1 testing.
   b. A nickel underplating is used below the tin.
   c. An insulation barrier of any kind is placed between the tin surface and adjacent conductor.

3. When the interconductor spacing is between 250 μm and 1 mm inclusive, tin may be used under any one of the following circumstances:
   a. The component lot has passed JESD 201 level 2 testing.
   b. A nickel underplating is used below the tin.
   c. An insulation barrier of any kind is placed between the tin surface and adjacent conductor.

4. When the interconductor spacing is less than 250 μm, tin may be used only if a solid insulator is interposed between the tin and the adjacent conductor. (Since this is impractical to achieve for fine-pitch surface mount devices, such devices are generally prohibited if they utilize pure tin finish.)

One advantage of the rules-based system is that it is simple to implement. The principal disadvantage is that it can only be applied to circumstances that were anticipated when the rules were written.

When novel packaging and design approaches appear that are not covered by the rules, adjustments will need to be made to the set of rules to accommodate these new realities.

The second methodology involves the use of a risk assessment algorithm, which is a form of expert system that has distilled the opinions of subject matter experts into a set of rules that are used to define a score. Risk assessment algorithms are also fairly easy to implement, although some training may be necessary in their use as with any tool. An advantage of an algorithm pool is that it provides a numerical figure of merit, which can sometimes be useful when more than a simple yes/no answer is needed. For instance, suppose tin has been inadvertently incorporated into a system and one needs "forgiveness" as opposed to "permission." It is important to understand not simply that a particular application falls into the region where tin would not normally be permitted; but one would like to know whether the application is just barely over the threshold or far into the region of high risk.

A tin whisker risk assessment algorithm has been developed by the author, and is available for download from a public website (www. reliabilityanalysisLab.com). However, other types of algorithms have been developed by users in the industry, and users are free to modify this algorithm or others, or develop their own as meets their needs.

Another means to assess the risk posed by tin whiskers for a given assembly incorporating tin-plated components is to use Monte Carlo techniques to simulate whisker growth within the assembly. This technique has an advantage of providing a quantitative risk of failure as a function of time. One major disadvantage of these techniques is that they require difficult to obtain knowledge of the whisker length, growth density, and growth kinetics probability densities. Such techniques are inherently more suitable for use by tin whisker subject matter experts than by purchasing agents or engineers who make minute-to-minute decisions on tin usage. However, these techniques can be very useful in the development and validation of design rules and other tools, and then they can be flowed out to the appropriate decision-makers.

A handful of researchers have utilized such Monte Carlo simulations, and some have even created simplified calculators that rely on Monte Carlo techniques to perform risk assessments. It is important to note, however, that the results produced by these models are based upon the properties of whisker fields that have been fully characterized by the researchers. However, given the variability of whisker growths and the difficulty of duplicating the whisker fields grown by the researchers, the actual risk posed by a particular application could be significantly higher or lower than that predicted by a simulation.

3.4.2   Tin/Bismuth Interaction with Tin/Lead

Tin/bismuth alloys are used both as a solderable finish and as the bulk solder material. Tin/bismuth solder alloys are much less commonly used for solder attachment than are tin/silver/copper or tin/copper alloys because of the potentially deleterious effects of the interaction between tin/bismuth and tin/lead. This concern becomes an issue when one of two conditions applies: either non-RoHS-compliant components with tin/lead finished terminations are assembled using tin/bismuth solder, or components with tin/bismuth finished terminations are assembled using tin/lead solder. Therefore, this concern relates to the back-compatibility of tin/bismuth components and the forward-compatibility of tin/lead components.

The lead/bismuth binary system exhibits a low-temperature eutectic at approximately 120°C. This creates the possibility that lead/bismuth alloys will suffer catastrophic mechanical breakdown as the temperature approaches 125°C which is close to the upper range of operational temperatures of many electronic systems. Because of this possibility, many manufacturers of high-reliability electronics forbid the use of tin/bismuth finishes on components in their products that they continue to manufacture using tin/lead solder assembly.

Whether or not the theoretical possibility of a low-melting-point phase actually affects a particular alloy will depend on its composition. The two different means by which tin/bismuth and tin/lead will come into contact with each other create very different resulting compositions. The amount of material that the component finish contributes to the overall solder joint will be far less than the amount of material contributed by the solder used to form the joint.

Most studies that have evaluated tin/lead finished components which have been soldered using tin/bismuth solder reveal that the long-term reliability of these joints is severely compromised with respect to thermal cycling. This is the primary reason why tin/bismuth solders did not become very popular.

However, most studies show that tin/bismuth finished components that have been soldered using tin/lead solder exhibit reliability equivalent to identical components that were finished with tin/lead. Good reliability seems to be maintained up to about 5 or 6 percent by weight bismuth content in the finish. Most tin/bismuth finishes that are deposited by electrodeposition contain approximately 3 percent by weight bismuth. There is a practical upper limit of about 4 percent by weight to the amount of bismuth that can be codeposited with the tin by standard plating practices. This factor, combined with the fact that bismuth costs more than tin, means that there is negligible risk that bismuth content of tin/bismuth finishes on components will inadvertently increase to high enough to create reliability concerns when used in combination with tin/lead solder.

Tin/bismuth alloy finishes are most commonly used on products manufactured in Japan. As of 2006, these finishes are used on approximately 5 percent of all the lead-free solderable terminations on components, although this fraction has been increasing over the past few years. The most common explanation for why some Japanese manufacturers prefer tin/bismuth over matte tin is that the addition of bismuth provides mitigation against tin whiskering. However, studies to date do not strongly support an assertion that electrodeposits of tin/bismuth exhibit a lower propensity for tin whiskering than do electrodeposits of matte tin.

3.4.3 Champagne Voids with Immersion Silver

The elimination of lead from tin/lead alloys used on PCBs has affected not only the solder interconnection compositions, but also solderable finishes for both PCBs and component leads. One promising lead-free circuit board finish that has gained popularity recently is immersion silver. Immersion silver is deposited by electroless process directly on to the copper surface of the circuit board runs. The resulting deposit is quite thin, on the order of 0.25 μm. The function of this very thin layer is simply to retard oxidation of the copper so that solderability is preserved during storage of the raw PCBs prior to assembly.

Under normal soldering conditions, such a thin layer of silver should be fully incorporated into the solder joint so that the resulting microstructure of the joint would be indistinguishable from a joint that was formed directly onto clean, bare copper. Occasionally, it has been discovered that some joints formed between SAC alloy solders and immersion silver exhibit a high degree of voiding along the interface between the solder in the copper. These voids appear as a layer of very small spheres, or bubbles. This appearance of numerous small bubbles has led to this phenomenon sometimes being referred to as *champagne voiding*, because it is reminiscent of the numerous tiny bubbles within champagne. The net result of this phenomenon is a drastic decrease in the robustness of the solder joints with respect to all forms of mechanical stress including thermal cycling life, shock and vibration resistance, and board flexure.

The manufacturers of chemicals and processes for performing immersion silver plating have investigated the root causes of the phenomenon, in an effort to remove the stigma of this concern from their product. The results of their work clearly show that the phenomenon relates directly to the conditions present during the immersion silver process. Groups of raw PCBs that are plated in a single lot should therefore be anticipated to exhibit similar levels of susceptibility to champagne voiding.

The mechanism that causes champagne voiding has been reported to result from the presence of very small voids in the copper immediately beneath the silver that result from the silver deposition process, which can be thought of as corrosion of the copper by the silver.

Manufacturers who wish to avoid the risk of introducing champagne voids into their solder joints should consider the following actions. It is recommended that electronics design and manufacturing engineers utilize immersion silver plating processes that have been provided by suppliers who demonstrate relative immunity to the phenomenon, and can provide proper validation data. End users do not always have the luxury of complete insight into the source of all plating materials and processes down to the PCB level. In this case, it may be necessary to evaluate each lot of boards for the propensity to form champagne voids in the course of the normal SAC SMT process. This evaluation will require cross-sectioning and will therefore be destructive.

The phenomenon of champagne voids again illustrates some of the risks associated with the implementation of a novel suite of materials. Soldering to immersion silver over copper was anticipated to be a very low-risk operation a priori. It was not until immersion silver was employed on a fairly widespread basis that this occasional plating process-related effect was noticed. After that, it required a few years of investigation to uncover the root causes so that appropriate mitigations could be employed.

3.5   Printed-Circuit Board Reliability Issues

During the initial investigations into the potential effects on reliability of the transition to green electronics, most of the attention was focused on solder joints. This seemed reasonable; because of the change to the alloy composition that would be required by the elimination of lead. However, the resulting selection of tin/silver/copper (SAC) as the most common new solder alloy created significant challenges for manufacturers of components and PCBs. These challenges result from the 40 to 50°C increase in temperatures required for reflow processes using SAC when compared with legacy tin/lead processes. As mentioned previously, these challenges are particularly acute when the requirements of rework are considered, because of the potential for multiple reflow cycles at the same location.

The principal concern from a reliability viewpoint is the degradation of PCB materials that do not result in immediate failure, but significantly degrade long-term performance of the printed-wiring board materials.

3.5.1   Crater Cracks

The first new failure mode identified that uniquely affects lead-free assemblies has been dubbed *crater cracks.* Intermittent connections were identified at corner interconnections between ball grid array packages and circuit boards that had been subjected to thermal excursions. This is symptomatic of solder joint failures as discussed

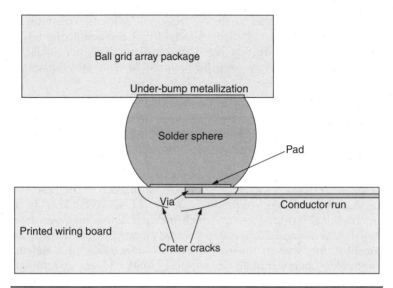

Figure 3.15 Schematic representation of a crater crack.

in the previous section, so initially was unsurprising. However, when these failure sites were analyzed, it was discovered that the solder joints were completely intact, and that the fracture had occurred within the circuit board itself. This fracture severed the connection between the circuit board runs and the pad at the corner location, causing the intermittent connection. This failure mechanism is illustrated in Fig. 3.15. In this example, the crater crack has created an intermittent open in an internal conductor trace that is attached to the corner solder sphere by a via.

The reason why lead-free assemblies are subject to this failure mode, while tin/lead assemblies are not, stems from the different constituent properties of the solder alloys. Tin/lead solder exhibits both a lower elastic modulus and lower yield strength than do SAC alloy solders. The higher stiffness and yield strength of the SAC alloy solders gives them a greater ability to transmit loads into the circuit board than is possible with tin/lead.

This new failure mode illustrates the complexity of the risks associated with the transition away from the legacy suite of materials, which have coevolved organically over a period of decades. Some may have naïvely assumed that because the new replacement material is "stronger" than the legacy material, reliability will improve. However, the discovery of crater cracks provides a cautionary example of how increasing the strength of one element of the system can have deleterious effects on other elements of the system, which then leads to an overall reduction in system reliability.

The solution to the crater cracking problem lies in modifying the circuit board materials so that they can withstand the increased loads that can be transmitted by the lead-free solder joints. This must be added to all the other challenges faced by PCB manufacturers who must produce boards that can withstand much higher soldering temperatures, and may soon be required to eliminate the use of halogens.

3.6   Aerospace and High-Performance Electronics Industry Standards

Shared concerns regarding the impact of lead-free technology on aerospace electronics have prompted the formation of the Lead-free in Aerospace Project—Working Group (LEAP-WG). This is an international working group which includes active members from North America and Europe. Represented among the membership are most of the world's major aircraft manufacturers and defense contractors, many middle-tier suppliers, and relevant government and customer organizations. The group has been working since early 2004 to develop a set of documents that provide guidelines and standard practices to meet the challenges of lead-free that are acceptable for use across our industry. These documents are being issued initially in the United States by the GEIA, and then they will be submitted to the International Electrotechnical Commission (IEC) for adoption globally.

The global electronics manufacturing industry is in the midst of the lead-free/RoHS revolution. Most commercial electronics manufacturers began delivering RoHS-compliant systems before the July 2006 deadline. The situation is very different within the aerospace and military electronics industries. Most of the products manufactured within these industries are excluded from the EU RoHS legislation or are covered by approved exemptions. Few, if any, aerospace and defense manufacturers have near-term plans to comply with RoHS. However, these manufacturers are already feeling the effects of RoHS and are working together to face challenges.

LEAP industries draw upon the same supply chain for electronic components and materials as do the commercial and consumer electronics manufacturers, and the volume of materials and components purchased by them is much smaller than that purchased by the latter. Not surprisingly, component suppliers focus their attention on the needs of their customers, most of whom demand RoHS compliance. Existing legacy products are sometimes supplied in two forms, but usually only temporarily before converting to the single RoHS-compliant version. New products are being introduced exclusively in RoHS-compliant (lead-free) form.

Avionics, military electronics, and other high-reliability electronics applications differ in significant ways from the vast majority of commercial and consumer electronics applications. Field environments often include extreme conditions: extreme climates, high altitude, high levels of shock and vibration, underwater exposure, and the extremes of space. Product lifetimes are often measured in decades rather than in years or months. Significantly, maintenance and repair activities are routinely performed down to the level of replacing individual components on PCBs. These maintenance and repair activities often occur many years after initial manufacture, at varied and distant locations, and under the control of agencies not always under control of the OEM. Finally, failure of the equipment to perform may have dire consequences.

Most of the principal stakeholders in the exempted industries realize the challenge posed by these issues. It is also recognized that consensus on common approaches will provide significant savings for the entire industry, as compared to the pursuit of divergent approaches.

3.6.1 Released Documents for the Aerospace and High-Performance Electronics Industry

1. GEIA-STD-0005-1, "Performance Standard for Aerospace and High Performance Electronic Systems Containing Lead-free Solder"

   This document specifies that users develop and implement written *lead-free control plans* (LFCPs). The purpose of the plan is to document processes that assure the plan owners, their customers, and all other stakeholders that aerospace and high-performance high-reliability electronics systems will continue to be reliable, safe, producible, affordable, and supportable.

2. GEIA-STD-0005-2, "Standard for Mitigating the Effects of Tin Whiskers in Aerospace and High Performance Electronic Systems"

   Although many aerospace electronics manufacturers will continue to use tin/lead as an attachment alloy for printed wiring assemblies, they will be forced to use piece parts with lead-free alloy finishes, the most common of which is pure tin. Pure tin finishes promote growth of tin whiskers, which can cause serious reliability problems in aerospace systems. The technical details of tin whisker growth and control are not completely understood, but their effects must be controlled in aerospace products. This standard provides a framework to execute certain levels of control and specifies that users develop and implement written tin whisker risk mitigation plans. Requirements for plans are structured according to standard levels of mitigation, which are selected by aerospace electronics manufacturers and users, based on the level of control required for the given application. Appendices to the standard provide guidance and insight into addressing risks associated with tin whiskers.

3. GEIA-HB-0005-1, "Program Management/Systems Engineering Guidelines for Managing the Transition to Lead-Free Electronics"

   This handbook provides assistance for programs in ensuring the performance, reliability, airworthiness, safety, and certifiability of product(s), in accordance with GEIA-STD-0005-1. Since the program manager is responsible for the overall reliability and performance of the product and since lead-free transitions may impact both reliability and performance, the purpose of this handbook is to illustrate what concerns should be voiced to ensure the lead-free transition does not have a negative impact on the product. This document was generated for both the program manager and the lead systems engineers who are responsible for ensuring that all system requirements are addressed and verified via design and integration.

4. GEIA-HB-0005-2, "Technical Guidelines for Aerospace and High Performance Electronic Systems Containing Lead-Free Solder and Finishes"

   This document provides technical guidance for the use of lead-free solder and mixed tin/lead/lead-free alloy systems while maintaining the high-reliability standards required for aerospace electronic and electrical systems. This document discusses such topics as (1) approach for analysis of tests and data, (2) lead-free solder behavior, (3) system-level service environments, (4) high-performance electronics testing, (5) solder joint reliability conditions, (6) components, (7) Printed-wiring boards, (8) Printed-wiring board assemblies, (9) module assembly conditions, (10) aerospace wiring conditions, (11) Repair and rework, and (12) Modeling and analysis.

3.6.2 Documents under Development

1. GEIA-STD-0005-3, "Reliability Testing for Aerospace and High Performance Electronics Containing Lead Free Solder" (planned date for initial release by GEIA: December 2007)

   The purpose of this document is to provide guidance for reliability testing of aerospace and high-performance products containing lead-free solder and a protocol for designing, conducting, and interpreting results from reliability tests. This document will provide a default method for performing reliability testing in the near term. Although several major reliability test programs are nearing completion, some time will be required before the data can be understood and characterized. In the meantime, manufacturers need a methodology to conduct their own reliability testing given their own unique set of service conditions. The protocol, presented in this document, is meant to be used when little or no other information is available to define, conduct, and interpret results from reliability tests for electronic equipment containing lead-free solder.

2. GEIA-HB-0005-3, "Guidelines for Repair and Rework of Lead-Free Assemblies Used in Aerospace and High-Performance Electronic Applications" (provisional title)

This document will be providing guidelines for repair and maintenance of lead-free electronics. This effort is just getting underway, so the anticipated release date has not yet been specified.

3. GEIA-HB-0005-4, "Guidelines for Performing Reliability Predictions for Lead-Free Assemblies Used in Aerospace and High-Performance Electronic Applications" (provisional number and title)

This document will describe methods of quantifying the effects of lead-free solder on system reliability and certification analysis. This effort is just getting underway, so the document title and number are not yet fixed, nor has the anticipated release date been specified.

3.7 Connector Issues

3.7.1 Cadmium Usage in Connectors

Connectors that are soldered onto printed-wiring boards are affected by the same challenges that all components face when exposed to the higher processing temperatures of SAC. In addition, some connector applications for harsh environments utilize cadmium plating to provide long-term corrosion protection and to maintain tribological properties on threads and other critical mechanical interfaces. Cadmium is on the list of RoHSrestricted materials. Although the use of cadmium in connectors is currently granted an exemption, how long this exemption will be provided is unknown. Furthermore, environmental regulations, other than the EU RoHS laws, currently restrict the use of cadmium, and new regulations under consideration also target cadmium usage. Therefore, it is likely that substitutions for cadmium-plated connectors will need to be employed on an increasing basis from now on.

Cadmium plating provides superior corrosion resistance for ferrous alloys that are exposed directly to the outdoor environment. The surface of the cadmium also exhibits low sliding friction and low contact resistance even the face of aggressive environments. This combination of properties makes cadmium an ideal choice, from the perspective of performance, for connectors that are subject to multiple mating cycles over an extended time while exposed to aggressive environments. As a result, cadmium plating has been incorporated into many military standards for use on a variety of connectors. In addition, cadmium provides a very close match to aluminum from the perspective of galvanic potential. It is therefore a good choice for finish when a ferrous metal must be mated against aluminum in a harsh environment.

Unfortunately, cadmium exhibits a high level of toxicity and is bioaccumulative. Thus, it is included on nearly every list of hazardous materials. The incorporation of cadmium-plated hardware into high-reliability systems creates a problem when the systems are decommissioned and must be disposed of.

No drop-in replacement has yet been found for cadmium plating on connectors that can replicate its superior performance. A variety of replacement options for cadmium have been considered, and are available. Some of these are listed in Table 3.13 together with advantages and disadvantages.

3.7.2 Tin Usage in Connectors

Another RoHS-compliant finish that has become popular on a variety of connectors is electrodeposited tin. The reliability concern with electrodeposited tin is the formation of tin whiskers (as discussed in previous sections). The use of tin on connectors poses a special problem from the perspective of tin whisker risk mitigation. Many of the mitigating techniques that OEMs can apply to other components, when a determination has been made the tin is not suitable, cannot be applied to connectors.

The application of conformal coating or other insulation is generally impossible, because the mating surfaces of the connectors must make electrical contact with one another. Dipping the component in tin/lead solder to replace the tin is also rarely possible with connectors due to their complex geometry and very demanding

Option	Comments
Stainless steel connectors	Increased cost, increased weight, limited availability, degradation of EMI performance
Nickel-plated aluminum	Decreased corrosion resistance
Gold plating	Increased cost, limited availability
Silver plating	Increased cost, limited availability, new toxicity issues

TABLE **3.13** Cadmium Replacement Options for Connectors

dimensional requirements. As result, OEMs that manufacture very high reliability equipment where tin cannot be used face a special challenge from tin-plated connectors. In fact, when all the components that are used in such high-reliability systems are considered together, connectors make up a disproportionate slice of the population of those components that pose the greatest difficulty when searching for nontin options.

3.8   Summary

The great change in electronic packaging materials and manufacturing processes that has been brought about by the EU RoHS regulation and other environmental regulations is historically unique. Previous changes have affected a small portion of the material suite and processes at a time, and have mainly been driven by the need for increased system performance. The current green electronics revolution is affecting a wide range of materials and processes, and the driving force is coming from outside the industry.

The potential effect of this revolution on the reliability of electronic systems has been extensively studied, modeled, and discussed. However, how this revolution plays out will not fully be known for several decades when the long-life systems currently under development reach the end of their design life.

There will inevitably be quality, and hence reliability, shortfalls as a result of the disruptions imposed by all these changes. Every new material and process has its own peculiarities, some of which will only be discovered through actual application. Champagne voiding and crater cracks serve as illustrative examples of the sorts of issues that can arise when new materials and processes are first implemented. It is reasonable to anticipate that such transient effects on system reliability will appear during the early years of our RoHS implementation. These problems are not likely to be distributed uniformly across all electronic systems, but will disproportionately affect small segments that have the poor luck to implement solutions beset by unanticipated failure modes.

Eventually, as the suite of materials stabilizes and lessons learned are incorporated to mitigate against these new challenges, the incidence of these transitional reliability issues should dissipate. A new baseline of anticipated reliability will become established. It is still an open question as to how this new baseline will compare with the legacy baseline that existed before our RoHS.

Acknowledgments

The chapter author wishes to acknowledge valuable assistance provided by Dr. Anthony Rafanelli, of Raytheon Integrated Defense Systems, for his insightful comments regarding the content of this chapter. I would also like to thank Forrest Fallier, Kenneth Arnold, and Mark Kostyla of Raytheon's Reliability Analysis Laboratory for providing many of the images used in this chapter. And finally, special thanks are in order for my wife, Susan, and my daughters for their patience on the many weekend days when they had to accommodate my monopolization of our library and its computer, while I worked on "the book."

References

Clech, J. "Acceleration Factors and Thermal Cycling Test Efficiency for Lead-Free Sn-Ag-Cu Assemblies." SMTA International, Chicago, September 25-29, 2005.

Fang, T., M. Osterman, and M. Pecht. "A Tin Whisker Risk Assessment Algorithm." 38th International Symposium on Microelectronics, Reliability I (Issues in Packaging), pp. 61-65, Philadelphia, Pa., September 25-29, 2005.

Galyon, G. "Annotated Tin Whisker Bibliography and Anthology." *IEEE Transactions on Electronics Packaging Manufacturing*, vol. 28, no. 1, January 2005.

Hamilton, C., P. Snugovsky, and M. Kelly. "A Study of Copper Dissolution During Lead-Free PTH Rework Using a Thermally Massive Test Vehicle." *SMTA International Conference Proceedings*, Chicago, September 2006.

Hillman, C., N. Blattau, E. Dodd, and E. Arnold. "Epidemiological Study on Sn-Ag-Cu Solder: Benchmarking Results From Accelerated Life Testing." *Proceedings of the 2006 SMTA International Conference*, Chicago, September 2006.

Hilty, R., and N. Corman. "Tin Whisker Reliability Assessment by Monte Carlo Simulation." *Proceedings of the IPC/JEDEC 8th International Conference on Lead Free Electronic Components and Assemblies*, San Jose, Calif., April 2005.

Kadesch, J. "Effects of Conformal Coat on Tin Whisker Growth." *NASA's EEE Links Newsletter*, March 2000.

Kadesch, J., and J. Brusse. "The Continuing Dangers of Tin Whiskers and Attempts to Control Them with Conformal Coat." *NASA's EEE Links Newsletter*, July 2001.

Kessel, K. "Joint DoD/NASA/Industry Lead Free Solder Project, Joint Test Report, Joint Council on Aging Aircraft/NASA Joint Group on Pollution Prevention." NASA Acquisition Pollution Prevention (AP2) Program, 2006. http://acqp2.nasa.gov/JTR.htm

Leidecker, H., and J. Kadesch. "Effects of Uralane Conformal Coating on Tin Whisker Growth." *Proceedings of IMAPS Nordic Annual Conference*, Helsingør, Denmark, pp. 108-116, September 2000.

Pan, N., et al. "An Acceleration Model for Sn-Ag-Cu Solder Joint Reliability Under Various Thermal Cycle Conditions." *SMTA International Conference Proceedings*, Chicago, September 2005.

Pinsky, D. "An Updated Application-Specific Tin Whisker Risk Assessment Algorithm—Within a Process Compliant to GEIA-STD-0005-2." *Proceedings of the 2007 Components for Military and Space Electronics Conference*, Los Angeles, Calif., March 2007.

Pinsky, D. "Enhancements to Current Tin Whisker Risk Assessment Methods." *Proceedings of the IPC/JEDEC 9th International Conference on Lead Free Electronic Components and Assemblies*, Boston, Mass., December 2005.

Pinsky, D. "Tin Whisker Application Specific Risk Assessment Algorithm." *Proceedings of the 2003 Military & Aerospace/Avionics COTS Conference*, Waltham, Mass., August 2003.

Pinsky, D., and E. Lambert. "Tin Whisker Risk Mitigation for High-Reliability Systems Integrators and Designers." *Proceedings of the IPC/JEDEC 5th International Conference on Lead Free Electronic Components and Assemblies*, San Jose, Calif., March 2004.

Pinsky, D., M. Osterman, and S. Ganesan. "Tin Whiskering Risk Factors." *IEEE Transactions on Components and Packaging Technologies*, vol. 27, issue 2, pp. 427-431, June 2004.

Pinsky, D., et al. "How the Aerospace Industry Is Facing the Lead-Free Challenge." White Paper: Prepared by the Lead-free in Aerospace Project—Working Group (LEAP-WG), December 2, 2006.

Touw, A. "Tin Whisker Interactions: A Monte Carlo Analysis." *Proceedings of the International Symposium on Tin Whiskers*, College Park, Md., April 2007.

Wickham, M., et al. "Measuring the Reliability of Electronics Assemblies During the Transition Period to Lead-Free Soldering." NPL Report DEPC MPR 030, National Physical Laboratory, The United Kingdom, August 2005.

Woodrow, T., and F. Ledbury. "Evaluation of Conformal Coatings as a Tin Whisker Risk Mitigation Strategy." *Proceedings of the IPC/JEDEC 8th International Conference on Lead Free Electronic Components and Assemblies*, San Jose, Calif., April 2005.

Industry Standards

JEDEC Standard JESD201, "Environmental Acceptance Requirements for Tin Whisker Susceptibility of Tin and Tin Alloy Surface Finishes."

IPC standard, IPC-9701, "Performance Test Methods and Qualification Requirements for Surface Mount Solder Attachments."

Government Electronics Industry Association Standard, GEIA-STD-0005-1, "Performance Standard for Aerospace and High Performance Electronic Systems Containing Lead-free Solder."

Government Electronics Industry Association Standard, GEIA-STD-0005-2, "Standard for Mitigating the Effects of Tin Whiskers in Aerospace and High Performance Electronic Systems."

International Tin Research Institute, ITRI Publ. No. 656, "Solder Alloy Data, Mechanical Properties of Solders and Solder Joints," 1986.

Websites

Center for Advanced Lifecycle Engineering, lead-free Home Page
http://www.calce.umd.edu/lead-free/
DFR Solutions Home Page
http://www.dfrsolutions.com/
iNEMI Lead-Free & Environmental Initiatives Home Page
http://www.inemi.org/cms/projects/ese/lf_hottopics.html
NASA Goddard tin whisker Home Page
http://nepp.nasa.gov/whisker/
NIST/SEMATECH e-Handbook of Statistical Methods
http://www.itl.nist.gov/div898/handbook/, October 1, 2007.
Raytheon Reliability Analysis Lab, tin whisker Web page
https://www.reliabilityanalysislab.com/TechnicalLibraryTinWhisker.asp
ReliaSoft's website
http://reliasoft.com/

CHAPTER 4

Environmental Compliance Strategy and Integration

Ken Degan

Teradyne Incorporated, North Reading, Massachusetts

4.1 Introduction

The emergence of the European Union Reduction of Hazardous Substances (EU RoHS) is one of many environmentally based regulations that impact the manufacturing of electronic products and subsequent business results. In addition to regulations, there are other drivers that shape an individual company's strategy for environmental and business success. Companies today are trying to balance the regulations, product quality, and customer value.

The intent of this chapter is to provide an insight on developing a strategy to meet the environmental requirements of today and tomorrow. The focus is on manufacturing strategies and the required infrastructure. Readers should use the examples and experiences described in this chapter and apply the strategies to meet their individual needs. Individual readers will have various levels of product requirements with both urgency to comply and complexity of products. The chapter will explore the spectrum of compliance requirements and how to shape an environmental compliance strategy. Once a strategy has been determined, it is essential to integrate and implement the strategy across all businesses within the company.

How is a successful environmental strategy born, what factors determine a good strategy, and how is it initiated? Like all successful

131

strategies, the directive must be supported from the top. The implementation of this strategy draws on key areas in the company and will compete with product development projects. Therefore, the strategy must communicate an imperative that is used to carve out key resources for the success of the compliance project. Some examples are presented later in this chapter that can be leveraged by readers to develop and integrate a successful strategy in their own business structure. A company needs to look at each of their products' market as it applies to the particular environmental regulation. The application of a regulation varies by products and market; the details will not be explored here, but it is key in shaping your company's environmental compliance strategy.

The methodology described here uses the backdrop of the EU RoHS directive (effective July 2006); however, principles and processes reviewed in this chapter could be applied to other environmental regulations. When considering environmental strategy, the market and products are key factors in how the regulations are applied and implemented. It is important to calibrate a baseline on the key factors to better apply it to your particular situation. To gain a frame of reference for the examples and experience explored in this chapter, readers need to be introduced to the author's perspective along with the company and the type of products where the synthesis of the chapter was realized.

Teradyne, Inc., designs, manufactures, and markets automatic test equipment (ATE) used by customers in multiple market segments. These ATE markets include semiconductor, assembly and systems, and automotive systems. The complexity of both the products and the markets served is diverse at Teradyne. At one end of the spectrum is the semiconductor ATE product line that can have over 100 printed-circuit board assemblies (PCBAs) of which some require liquid cooling with system accuracy measured in picoseconds (ps) and nanoamperes (nA). At the other end of the spectrum is the handheld equipment for testing automotive systems used to diagnose car problems.

Teradyne's insight on fundamentals and position of environmental stewardship is presented in an excerpt from its external website statement intended for customers, employees, and suppliers.

Teradyne is committed to promoting, creating, and maintaining a safe and healthy workplace and to improving the environmental quality of our operations and surrounding communities. This effort begins with providing a safe physical plant and hazard-free working conditions. To accomplish these goals, Teradyne adheres to the following Environmental & Safety (E&S) principles:

- Minimize any significant adverse environmental impacts or safety/health risks to employees, customers and the public, through the use of integrated E&S management procedures and planning;

- Maintain compliance with all applicable international, federal, state and local regulations and laws;
- Prevent pollution, reduce waste emissions, and commit to reuse, recovery and recycling;
- Secure and maintain all applicable permits and other regulatory approvals required for operations; and
- Continually assess and anticipate future E&S laws and regulations and the effects they may have on our operations.
- Teradyne is committed to continuous improvement in all activities with focus on pollution prevention, waste minimization and reduction in use of chemical raw materials. We work to improve our products and processes to reduce any negative impact on the environment, safety and health.

It is this type of corporate backdrop that plays an important role in the success of a compliance project initiative. How your strategy is developed and executed is largely based on your company's overall corporate position and past performance of environmental directives.

4.2 Describing the Imperative

To better understand what drives the process, it is essential to define and communicate the potential impact on the product for corporate management. Here, the issue of concern is the EU RoHs directive that banned six substances—lead, mercury, cadmium, chromium poly-brominated biphenyls (PBBs), and polybrominated diphenyl ethers (PBDEs)—commonly used in today's electronics. Of these six substances, lead removal can have the greatest impact on PCBA processes and therefore electronic products.

The industry has identified major technical gaps in using lead-free processing in many product applications. Identifying the potential impact that these gaps could have on your company's products is essential in describing the effort needed to drive products to meet environmental regulations compliance. Areas of impact could include interruption in revenue, product reliability, and penalties for not meeting the particular environmental requirement. The next step is the development of a technology road map to solve these gaps that will keep your company from meeting environmental regulatory compliance. The imperative will most likely be driven by the operations group (manufacturing) in identifying a discontinuity in the manufacturing and attachment road map and by the corporate compliance group, both of which are on the frontlines of the dynamic environmental landscape. It is worth noting that the imperative is not driven from marketing or product design engineering where most product initiatives are traditionally generated.

To further this discussion, let's look at an extreme or complex product example where the product attributes would cause a discontinuity in the current manufacturing road map. Thermal gradients and peak temperatures seen by PCBAs with varying component thermal attributes, in conjunction with a thick printed-circuit board (PCB) during lead-free reflow, can be extreme across the PCBAs. Furthering the understanding of complex PCBAs, below are manufacturing and PCB attributes that would encompass complex PCBAs:

- Over 60,000 Solder Joints
- Over 13,000 placements
- Over 80,000 opportunities for defects

These factors along with the increased temperatures needed for lead-free processing can cause a negative impact in the application of complex PCBAs. Technology discontinuity for lead–free processing must be overcome before a company can ensure customers that lead-free processing does not negatively impact product reliability. Along with the discontinuity of the road map, survivability of components during lead-free reflow is also in question. Describing the imperative is essential in bringing to bear the right resources to meet new environmental regulatory requirements.

4.3   Constructing the Team

Forming a comprehensive team involves identifying the product line and the individual manufacturing technology requirements along with team members who understand the market and environmental regulatory impacts. The team members may include representatives from multiple functions in the organization. Representatives from product development, corporate environmental compliance, multiple functions in operations (including manufacturing engineering, test engineering, and supply chain), and project management support. Team formation may vary depending on product and business complexity. Based on product complexity and compliance time line requirements, the structure could be a cross-functional full-time team (heavyweight team) or cross-functional part-time team (lightweight team). Regardless of whether the team structure is heavyweight or lightweight, team commitment and the requisite skills and functions are vital.

Figure 4.1 shows an example of a cross-functional team structure, which would enable a team to draw upon skills companywide under the direction of a project leader and a project manager. Here, the team structure is depicted as a lightweight team with a full-time project manager. Determination of what part of the organization the project leader would come from would be based on the particular technical and business challenges. Complex products may be led by product development. In other situations, it could be led by operations or the

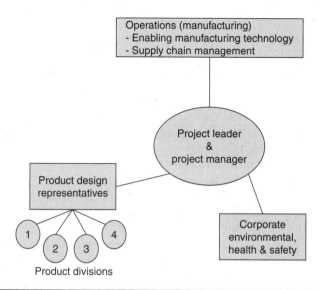

FIGURE 4.1   Example of a team structure.

corporate compliance functions. It is important that the product design representatives have broad knowledge and credibility to drive key product initiatives for this type of multifunctional organization (sales, marketing, design engineering, artwork design, and product planning). The core members in the example in Fig. 4.1 totaled nine, but depending on a company's products or the organization's complexity, this number would vary.

A vital element for strategy development projects with this complexity is a steering or management committee. The committee oversees and supports the team and enables the team to maintain its focus on the initiatives. The role of the committee also would include budget allocation, resource allocation, strategy approval, management changes, and project implementation. It is important to establish a management body such as the management committee that will be able to provide support for the initiatives impacting environmental compliance throughout the company. Major change in product or practices must be a top-down initiative in order for it to be a successful strategy.

The team makeup is key in providing the correct skill and function required for the particular product complexity and business requirements. What are the functions needed in a team addressing complex products? Using the team example discussed in Fig. 4.1, the roles and responsibilities can be further detailed.

Project leader. The project leader provides technical leadership and gains corporate support. In addition, the project leader is responsible for developing and executing the strategy. The project leader is also required to get budget approval and meet budgetary constraints.

Project manager. The project manager is responsible for following project management processes and developing and maintaining a project schedule. The project manager provides process and drives the project initiatives.

Corporate compliance. Corporate compliance is responsible for understanding and gathering environmental regulation requirements worldwide and advocating the company's product and business impact on those regulations.

Supply chain management. Supply chain management is responsible for supplier compliance road maps and the company material declaration component attribute database.

Manufacturing engineering. Manufacturing engineering is responsible for identifying and mitigating technology gaps to enable the products to meet customer and business requirements. The roles include manufacturing and engineering.

Product design representatives. The product design representatives are responsible for having a solid understanding of product, customer, and market requirements.

Now with the team formation and structure determined, strategy planning can be initiated. It is not important for us to go into detail about the planning process, but to recognize it as essential for outlining and implementing an effective strategy. Having a formal process allows for a common language and planning deliverables to drive complex initiatives. A revolutionary product development (RPD) framework represents an example of a formal planning process used for product development.

RPD framework uses phase gates. Phase gate 2 of the process is where detailed planning is developed. Below is some generic collateral that could be used in a detailed plan development.

4.3.1 Planning Process Outputs

- *Roles and responsibilities.* Development of a responsibility matrix requires credible resource project alignment and to gain origination commitment for the resource. The roles and responsibilities should also include both duration and effort for each role on the project. This is especially important for a lightweight team structure.

- *Scope.* Scope should include the detailed description of the project and why the project must be done. In addition, project deliverable and quantifiable success criteria must be included. Example: "Product X must be RoHS-compliant by July 2006."

- *Project schedule and budget.* This includes key schedule and budget plans (both internal and external). Conduct a critical path analysis to better understand project constraints. This will allow for better resource planning during the initial project planning where it is more effective to gain resources in flight.

- *Change control.* Describe the process including approval of changes with schedule resource or deliverables of the project. This is required to avoid scope creep without the trade-off analysis.

- *Management review.* Describe the cadence or project events that will necessitate a management review.

- *Risk management.* Process to identify, determine impact, and develop mitigation plan for project risk that emerges during the project. Risk should be managed in key categories, e.g., schedule, budget, supplier, customer, and technology.

4.4 Developing the Strategy

Every company needs to assess its products against the particular compliance directive to see how the directive is applied. The development of the strategy should take into account product complexity, technology gaps, supply chain, compliance road maps, and regulation exemptions and exclusions along with market drivers.

Examples of major strategy points for EU RoHS that should be included are as follows:

1. Meet the requirement of the law or directive.
   a. Meet all environmental regulations as they apply to the company's products.
   b. Perform due diligence in determining how the environmental directive pertains to your product and market.

2. Transition the products to meet compliance.

3. Design new products using EU RoHS-compliant materials and migrate toward lead-free processing as technology gaps are mitigated.
   a. Establish a *design for environment* (DfE) process for all new products.

4. Describe the importance of the customer considerations in the strategy.

5. Defer conversion for EU RoHS-compliant (lead-free) products until lead-free processing does not negatively impact product reliability.

6. Continue to drive technology requirements to meet compliance.

Even if your company's products are out of scope of the particular directive, the strategy should account for major drivers such as market, supply chain, and regulatory changes that may impact the strategy. All the while, the strategy should drive mitigation for identified technology gaps that would affect product reliability, which would impact the customer.

4.5   Gaining Approval of the Strategy

With the team and management review structure established and the detailed planning complete, now is the time to get your strategy approved. Here are some important points to consider when getting your strategy approved.

- Review the strategy and initial plan including resource and budget requirements.
- Outline preliminary schedule with major milestones identified.
- Break down reporting schedule and budget performance.
- Partition project deliverables into tracks or track teams to better utilize skilled resources.
- Manage interdependencies of the tracks at the project level.
- Maintain regular cadence with management committee meetings to support and report progress of the strategy implementation.
- Continue to monitor the key drivers that may impact the strategy going forward.

It may take more than an initial committee meeting to approve budget and resource allocation. Depending on the company, significant preparation work may need to be done to bring corporate and division management up to speed on the significance of the compliance directive. Once the approval is set and the team identified, it is time to execute.

4.6   Launching the Initiative

With the strategy developed and the planning complete, the initiative can be launched. Once you have organized the team and appointed a project manager, the next step is to outline a meeting schedule for communication and to drive the newly formed project plan. It is helpful to develop a project plan in multiple tracks. For example, the tracks could be defined as manufacturing technology, supply chain, regulatory, and product design. Depending on the location of your team members, it is key to outline a communication schedule to regularly update members on progress and results. Companies with products that have a wide spectrum of technology or market may find that team members are geographically distributed or have had minimal interaction with one another in the past. If this is the case, scheduling face-to-face interaction with such cross-functional teams can help cultivate the communication process.

4.6.1   Communicating Companywide

With such diverse market and product considerations, the team must focus on a companywide communication plan to avoid having this become an obstacle in the planning process or the strategy implementation. A key objective of a project is to communicate the strategy. One way to do this would be to develop awareness presentations for review with such groups as product design, operations, field service, and key supply partners. Another suggested format would be tap into already scheduled technology exchanges to communicate. The use of an internal Web page broadens the ability to parlay the strategy companywide. Reassuring customers about the directive can be achieved through a targeted external website campaign.

Here is an example of an external website that was used by Teradyne Inc. to communicate its EU RoHS strategy.

Teradyne's Global RoHS Mission Statement

Teradyne is committed to complying with all applicable laws and regulations, including the European Union Restriction of Hazardous Substances (RoHS) directive that restricts the use of hazardous materials in electronics products. Teradyne continues to exceed RoHS directive compliance obligations on a global basis. The company continues to work toward the reduction of RoHS materials in its products which are subject to the RoHS directive, except where it is widely recognized that there is no technically feasible alternative. Teradyne has been working on RoHS lead–free initiatives since the mid-1990s and participates in several groups looking for alternatives to RoHS substances and striving to harmonize the RoHS standards globally.

In 2004, Teradyne established RoHS project teams to assess RoHS implementation alternatives and to understand the supply line issues associated with conformance to the directive. These teams are working with Teradyne's suppliers to ensure that RoHS-compliant materials do not adversely impact the quality and reliability of Teradyne's products. In addition, these teams oversee several complex engineering projects to ensure that our manufacturing processes meet our design for the environment initiative.

It is important that our employees, customers, and suppliers know that Teradyne continues in its strong commitment to protecting and improving the environment for future generations. Our work on RoHS is just one of the ways in which we strive to be an industry leader in demonstrating stewardship for our environment.

To communicate effectively, the team should consider presenting multilevel communication packages. For groups such as design engineering, presentations will be more technical in nature, encompassing the impact of design and technology gaps for lead-free products. To address operations and supply chain management groups, workshops could be designed to communicate and target their specific areas of impact such as PCBA issues and supply chain conversion trends. Some other areas that should be addressed are product reliability and component compatibility with lead-free processing. A clear technology road map is important in communicating technology gaps and resulting in gap mitigation. It is worthwhile to keep in mind that lead is one of the six banned substances called out in the EU RoHS directive.

For a team to maintain a clear focus, it is important to identify and establish key drivers throughout the project. Key drivers that should be monitored are supply chain, market, regulatory compliance, and product design as influenced by DfE.

Figure 4.2 represents the drivers as they relate to the strategy. This could also be used to articulate current status and the trend of the risk drivers. This figure could be helpful in addressing any concerns that may arise over the forces that may impact the current strategy. In the diagram, a stoplight assessment can be used in the status/trend boxes. This can allow for better communication with groups such as the management committee by enabling them to visualize areas that may need to be explored while continuing to validate the strategy.

Once the team has been identified, the risk drivers should be monitored for both the current status and what the trends appeared to be. In doing so, a company can continually assess the strategy to make sure it is meeting dynamic changes in the market and industry.

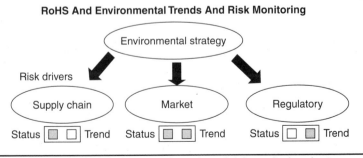

RoHS And Environmental Trends And Risk Monitoring

FIGURE 4.2   RoHS and environmental trends and risk monitoring chart.

4.7   Driving the Project

As discussed earlier, managing the project in tracks is recommended. To better illustrate the advantage of track management, multiple potential tracks are explored here in greater depth. These tracks may be different for every company depending on market and product complexity. Some examples of tracks that may be considered include manufacturing technology, supply chain, regulatory, market, and product design (Design for Environment [DfE]).

- Manufacturing technology track assesses and mitigates gaps moving to a lead-free process and staying in leaded process during product transition.

- Supply chain track addresses component information procurement and infrastructure required to move toward a EU RoHS-compliant product.

- Regulatory track drives due diligence to determine product inclusion or exclusion from the EU RoHS directive.

- Market drive track monitors and assesses dynamic customer and market requirements.

- Product design track develops processes to use compliant material in new products.

Some of the tracks will be referenced further in this chapter as examples only. Based on your particular company's strategy and products, the tracks may differ.

4.7.1   Manufacturing Technology Track

To assess technical gaps, a compliance team should use a combination of industry benchmarking data along with design of experiments (DoE) to understand lead-free manufacturing impacts on products. To better leverage industry knowledge, it is recommended to participate in industry groups and consortiums.

To gain a better comprehension of the impact of a lead-free process, consider converting a current product to a lead-free product and conducting reliability testing. This will provide quantified data to help determine lead-free processing technology gaps and to mitigate the impact on reliability and manufacturing yield. Benchmarking and a series of design of experiments will highlight key technology gaps that must be overcome to achieve desired product reliability. Figure 4.3 summarizes a method to track these technology gaps as they apply to your products. The areas noted in Fig. 4.3 of greatest concern are lead-free PCB process survivability and device reliability due to required temperature increases during initial reflow and rework profiles. Rework is essential in large complex PCBAs, due to the high value of their components.

Risk by Commodity	Externally Driven	Internally Driven
Passive components	ELECTROLYTIC & FILM CAPS	SMT & PTH CONNECTORS
Semiconductor components	QFP, QFN PACKAGES	BALL GRID ARRAY (BGA) PACKAGES
	THERMAL SLUG DPAK SOIC & QFN	CBGA/CCGA PACKAGES
Specialty components	RF/MICROWAVE COMPONENTS	OPTOFET & REED RELAYS
PCB technology	GENERIC HIGH-TEMPERATURE FR4 LAMINATE	LOW DK/DF LAMINATE
Manufacturing and rework processes	PTH REWORK	MIXED/LEAD-FREE MASS REFLOW
		MIXED/LEAD-FREE BGA REWORK

FIGURE 4.3 Technology dashboard of the manufacturing issues that may be tracked by a team.

Figure 4.3 is a summary of technology risks that can be identified by multiple means and sources including DoE of converting a company's product(s). The DoE would include yield monitoring using IPC-A-610 and reliability testing. All risks identified would have to be addressed before products can be converted to lead–free. The drivers identified in this example are all considered a high concern for a company with complex products. Technology risks can be divided into two categories: One is driven by many companies that would need to solve the problem; these risks would be driven externally. The other risks that are unique to a company's product requirements would be driven internally. In focusing on the product risks, it is important to look at various commodities and processes including passive components, semiconductor components, specialty components, bare board technology, and manufacturing and rework processes as compared to your product.

The sources that can be exploited for the risk identification are:

- Contract manufacturers' lead-free road map
- Consortiums that include companies that share regulatory requirements and manufacturing capability requirements

FIGURE **4.4**   Delimitation of PCB due to elevated temperature in solder reflow.

As seen in the previous photographs (Figs. 4.4 and 4.5), failure of both attachments and massive PCBs occurred based on the increased temperature of reflow. With this type of information, a team could launch multiple technology projects to mitigate these technology gaps. Figure 4.6 can be used to determine relative risk in transitioning to lead-free based on product attributes.

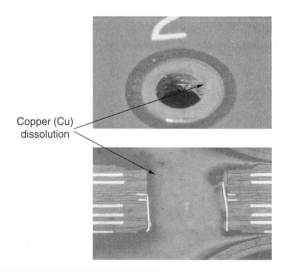

FIGURE **4.5**   Copper dissolution from plated through-hole (PTH) to solder caused by required higher-temperature processing.

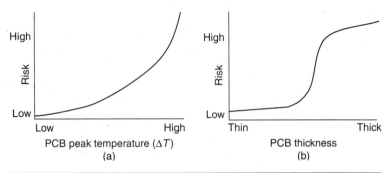

Figure 4.6 Chart showing relative risk based on technology attributes (PCB peak temperature ∆T and PCB thickness).

One example of a technology gap that would need to be driven internally is the qualification of high-performance PCB laminates for lead-free manufacturing as applied to products both for performance and for lead-free processing survivability. Working with laminate suppliers directly along with PCB fabricators, the team could design a test vehicle that had the required attributes for the product PCB applications. In this example, the test vehicle laminate tests included the following.

Laminate Tests

- Dielectric constant D_k
- Dissipation factor (Df)
- Time to delamination at 260°C (T260)
- Time to delamination at 288°C (T288)
- Glass transition temperature T_g
- Thermal decomposition temperature T_d
- Copper peel strength

Reliability Tests

- Interconnect stress test (IST)
- Highly accelerated thermal shock (HATS)
- Conductive anodic filament (CAF)
- Solder float with micro-sections

Figure 4.7 is a summary of testing performed to determine PCB laminate compatibility lead-free process (reflow elevated temperatures). In addition, refer to Chap. 8 for more details on laminate testing.

Here, the team was able to develop a road map for a successful technology solution partnering closely with technology leaders in the PCB and PCB material industry. Other projects included investigating

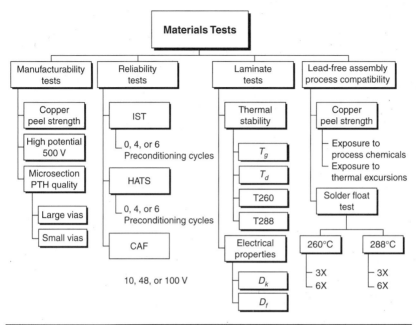

Figure 4.7 Summary of Testing performed to determine PCB laminates compatibility lead-free (Pb-free) process (re-flow elevated temperatures).

rework capability and key device lead finishes and thermal stress that may impact products.

The manufacturing track should deliver a comprehensive technology road map that would enable capability or mitigate technology gaps impacting the ability of products to comply with environmental regulations without impacting the customer. Key triggers should be developed to ensure capability meets the dynamic compliance time line.

4.7.2   Supply Chain Track

This track deals with the lack of EU RoHS compliant information in the information database. This is required to make essential design and component sourcing decisions. Here are two common issues regarding supply chain infrastructure that arise when moving a product to meet compliance requirements.

1. Lack of relevant component data to support a company's environmental compliance product analysis and to support its product compliance road map

2. Lack of component and product attribute database infrastructure

When looking at infrastructure solutions, a team should consider both short-term and long-term solutions. This will allow the project to move on without the full burden of major infrastructure change. As commonly known, infrastructure change requires major resources and time. The long-term infrastructure change can be done in parallel with the project. A short-term solution may include providing data for soldered components and providing component information to influence new product part selection (compliant and lead–free process capability information). A long-term solution for the product attribute database infrastructure would include detailed material declaration and being more scalable so as to manage the dynamic environmental compliance requirements.

A supply chain concern that may arise is the supplier discontinuing components compatible with leaded assembly process during your product transitions. Ball grid arrays (BGAs) are not process-blind, meaning they are not forward and/or backward assembly process compatible. BGAs using (tin/silver/copper) SAC solder are not compatible with a leaded process. One commodity that converted over to lead-free aggressively was BGA memory devices. Market segments that were required to produce compliant products drove these devices. So as to not impact the revenue stream, a solution would need to be developed and implemented to mitigate risk of the supply chain conversion from lead to lead-free for BGA memory devices. The following is an example project to mitigate this transition by providing manufacturing capability to use SAC in a leaded assembly process.

The project considered three options to mitigate this risk.

1. Replace BGA SAC solder balls with tin/lead balls and process PCBAs with leaded reflow profiles.

2. Place and reflow BGAs with SAC using a BGA rework process after PCBA mass reflow. This would allow only the BGAs to experience the elevated temperature required for lead-free processing.

3. Develop mass reflow capability for mixed alloy soldering (SAC solder balls and tin/lead solder paste). Slightly elevate reflow temperatures, producing reliable mixed alloy solder joints.

In this case, it was determined that option 3 was the best approach. A mixed alloy reflow profile was developed by increasing the peak temperature by less than 10C of the SAC balls with the tin/lead solder paste. (Failure to fully melt and mix SAC balls with tin/lead solder can produce solder fillets that fail prematurely.) This solution enabled the use of lead-free BGAs in combination without placing other thermally sensitive components at risk.

Figure 4.8 refers to a reflow profile chart highlighting the comparison of reflow temperatures. The gray area of the curve depicts the reflow temperature that allows for proper mixing of the two alloys. Please consult sections 6.4.7 and 7.3.2 for more details on this

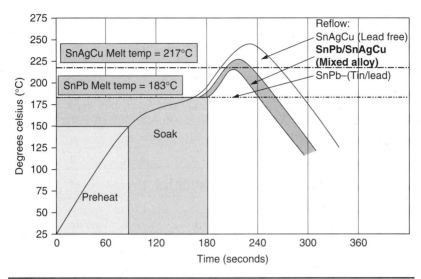

FIGURE 4.8 Reflow temperature profiles for lead versus lead-free process along with the mixed solder alloy reflow profile.

problem based on similar experiences in Benchmark Electronics and Tyco Electronics.

Depending on the product, design of experiments (DoE) should be used to determine the best course of action needed to manage supply chain conversion to lead-free. It is highly dependent on PCB complexity (thickness).

The supply chain track will deal with internal infrastructure changes and the unpredictability of the supply chain conversion. This will challenge both the current products' end-of-life strategy and the new products' conversion strategy.

4.7.3 Product Design (Design for Environment) Track

This track would develop a DfE strategy to apply to new products. This will incorporate lessons learned from the manufacturing track along with the resulting technology road map. Considerations of the product and product development process must be used to develop the DfE strategy.

Modification of CAD libraries to alter land pattern geometries to be compatible with lead-free process must also be addressed.

4.7.4 Market and Environmental Regulations Track

Considerations of individual customers, global markets, and companies' competitive landscapes are vital in shaping the company's strategy for environmental compliance. A process for monitoring these market drivers is essential to the success of the strategy. As indicated earlier in the chapter, environmental regulation impact varies according to the product and market served. But common areas include providing due diligence of position of compliance, along with strong networking with product-compatible industries and advocating companies' product compliance position with governing environmental regulatory agencies.

Working Definition of Due Diligence   Take all reasonable steps and exercise care that a reasonable person exercises under the circumstances to meet the requirement.

4.8 Reflecting on the 132Environmental Compliance Strategy

Companies with varied product portfolios have to consider the consequences of the regulations and how they affect product lines differently from full compliance required to product exclusion in the case of EU RoHS. When a product is in scope of the regulation, a combination of taking the product off the market and redesigning products must be undertaken. This action would require close communication with affected customers to continue to provide alignment with customer's requirements.

Even products that are exempt from the EU RoHS directive will still require monitoring of the key drivers along with a road map of product compliance. It is recommended to employ a time line for these products to transition to lead-free designs that, in turn, can drive all new product development to use RoHS-compliant components. The knowledge gained here can be used to publish an internal design for environment guideline and provide product training to design engineering groups.

Technology gaps may limit compliance in some products. Timely conversion of RoHS-compliant material will reduce future product redesign when the manufacturing technology gaps are implemented. Key triggers should be developed with required runways to meet changing regulatory requirements that could impact product revenue. It is important that you continue vigilance and keep the team together to monitor the dynamic worldwide regulatory changes and drive technology gaps to meet future environmental compliance requirements.

4.9 Conclusions

Environmental requirements facing today's companies are challenging. The purpose of this chapter was to present strategies to meet the dynamic environmental requirements of today's PCBA manufacturing. The reader should consider the examples presented here as sample strategies that can be adapted to individual situations.

This chapter focused on incorporating regulatory compliance, product quality, and customer value. By breaking down the strategy in steps from outlining the team to launching the initiative, the reader can see clearly each step and what is needed to shape and develop an environmental compliance strategy. With a strategy determined, the next series of steps includes integrating and implementing the strategy, all the while communicating the strategy across all businesses within the company. Obviously, the urgency level of a company to comply will increase the intensity, but this chapter shows how to establish a systematic process using track management. The tracks explored here were manufacturing technology, supply chain, regulatory, market, and product design DfE, but these may vary depending on a company's market and product complexity. Each of the tracks drives such issues as the impacts of lead-free processing, technology gaps that have to be addressed for products to comply, the lack of industry compliant information, and the drivers that contribute to a company's compliance schedule. Using the track management approach is a successful outline for achieving environmental compliance. The goal throughout this chapter has been to give the reader a process by which to work toward environmental compliance without impacting product reliability.

Acknowledgments

The author wishes to acknowledge the members of Teradyne, Inc., RoHS Integration and Strategy Project (RISP) Team including Kenneth Craig, David Evans, Garry Grzelak, Kevin Hughes, Dane Krampitz, Jonathan Meltzer, Debra Pulpi, Kathy Queirolo, Joyce Roberson, and Dan Wiesen. In addition, the author wishes to thank Valerie St. Cyr of Teradyne, Inc., for technical contribution, and Pamela Degan for providing editing support. Finally, thanks to all the members of the New England Lead Free Consortium.

References

Clark, K., and S. Wheelwright. *Revolutionizing Product Development: Quantum Leaps in Speed, Efficiency, and Quality*. New York: Free Press, a division of Macmillan, Inc., 1992.

Teradyne Incorporated. October 2007 external website http://www.teradyne.com/corp/env_pol.html.

St. Cyr, Valerie. "New Laminates for High Reliability." *Proceedings of IPC Printed Circuit Expo*, Anaheim, Calif., February 2006.

CHAPTER **5**

Managing the Global Design Team in Compliance with Green Design and Manufacturing

Mark Quealy

PMP, Schneider Electric, North Andover, Massachusetts

5.1 Introduction to Implementing Green Design for Electronic Products with Global Teams

We live in a disposable society driven by electronic technology, resulting in huge amounts of obsolete and discarded products which may find themselves in landfills, affecting regional water supplies. Recently initiated by government environmental directives, efforts are underway to change the way we design, manufacture, and discard these items.

Managing the implementation of environmental directives poses a considerable challenge to most organizations. On the surface, it appears to be a good idea, designing in a safe and sustainable way, offering quality products that meet customer needs, that do not waste resources, and that do not have a negative impact on the environment. However, once the scope of compliance is apparent, what one is being asked to accomplish while managing the business is overwhelming: how is one to comply with all regulations while continuing to launch new product development projects with fewer people, less time, and utilizing global teams.

Any new project will involve practically all the disciplines within the company, the executive staff, business directors, marketing, engineering, manufacturing, quality, purchasing, and logistics. In most cases this compliance expenditure in time, resources, and money does not add value to the product, or any new features, and could possibly increase the cost of goods sold, impacting the bottom line.

Environmental directives pose difficult design issues as proven designs are altered with new parts and materials, creating many technical problems. It goes against the tendency of "if it is not broken, don't try to fix it" to let the product run its life cycle and end its commercialization.

These directives require investment in new or retrofitted capital equipment for the factory. Considerable effort will be spent to put the same product, form, fit, and function back in stock as before. But this is the responsible action for any company as good corporate governance and a citizen of the world. According to the *Financial Times* issue of September 19, 2006, "A key challenge is to become ready for all types of compliance rather than addressing pieces of legislation as they are introduced, not being reactive and inefficient!"

The customers' perception of a company becoming a green supplier can be used as a competitive advantage, and that sometimes offsets potentially negative cost increases. In this chapter, the aspects of planning and implementation of converting products to more environmentally friendly forms are discussed as well as the design project challenges that arise as a result. The goal of this chapter is to assist any company in the successful implementation and adoption of green design practices.

5.2   Understanding the Scope of the European Union Environmental Directives

The European Union (EU) is an integration of economies throughout Europe. Member states surrender a certain amount of their individual sovereignty to manage common economic, social, labor, and environmental health and safety issues. From this collaboration, they can gain the greatest impact and benefit. The Council of Ministers was established as a legislative body to create regulations, policies, and directives and ensure smooth implementation throughout the EU. The most recent directives affecting the electronics industry have originated in the EU and are the Restriction of Hazardous Substances, 2002/95/EC (RoHS); the Waste , Electrical, and Electronic Equipment, 2002/96/EC (WEEE); and the Electromagnetic Compatibility (EMC) Directive, EN61131-2 2003. Given these three directives, the objective of a green company will be to redesign existing products as well as implement

Banned substances covered by the restriction of hazardous substances	Maximum levels (ppm)
• Cadmium (Cd)	100
• Mercury (Hg)	100
• Lead (Pb)	1000
• Polybrominated biphenyls (PBBs)	1000
• Polybrominated diphenyl ethers (PBDEs)	1000
• Hexavalent chromium (CrVI)	1000

FIGURE 5.1   RoHS banned substances.

new tools and rules into the products' development processes. These websites are recommended to become familiar with RoHS:

http://ec.europa.eu/environment/waste/weee/index_en.htm
http://www.rohs.gov.uk./

5.2.1   The Restriction of Hazardous Substances (RoHS) Directive

The RoHS directive bans mercury, lead, cadmium, hexavalent chromium, and two brominated flame retardants, PBB and PBDE, in electrical and electronic equipment unless specifically exempted. The list of banned substances and their target levels is shown in Fig. 5.1

The objective of this legislation along with the WEEE directive is to minimize the impact of electronic manufacturing, use, treatment, and disposal of products on human health and the well-being of the environment, that is, to develop environmentally benign design materials and manufacturing processes. The root cause of this issue is due to the disposal of electronic waste in landfills and leaching into the area groundwater supplies which eventually may travel to municipal water systems and affect human health.

RoHS requires the analysis of existing materials, electronic components, and hardware for the banned substances and designing in replacements. Specifically, all new lead-free materials, while compliant with regard to their compositions, have the additional burden of being able to withstand elevated manufacturing process temperatures. Additionally RoHS efforts will be adversely affected by the obsolescence of current noncompliant components by manufacturers who have decided that the cost of conversion is too high for the remaining market that exists.

5.2.2   Understanding Affected Product Categories

The following product categories are impacted under the RoHS directive:

1. Large household appliances: refrigerators, washers, stoves, air conditioners

2. Small household appliances: vacuum cleaners, hair dryers, coffee makers, irons, etc.

3. Computing and communications equipment: computers, printers, copiers, phones

4. Consumer electronics: TVs, DVD players, stereos, video cameras

5. Lighting: lamps, lighting fixtures, lightbulbs

6. Power tools: drills, saws, nail guns, sprayers, lathes, trimmers, blowers

7. Toys and sports equipment: videogames, electric trains, treadmills

8. Automatic dispensers: vending machines, ATMs

5.2.3   Currently Exempted Product Categories

1. Large stationary industrial tools

2. Control and monitoring equipment

3. National security use and military equipment

4. Medical devices

5. Some lightbulbs and some batteries

6. Spare parts for electronic equipment in the market before July 1, 2006

There is no requirement for marking or labeling of products which comply with the RoHS directive; it is up to the manufacturer if they want to use an identifier. Some may display a marking such as the one highlighted in JEDEC standard JESD97 shown in Fig. 5.2. Other more colorful and graphic markings are available through label vendors if desired.

5.2.4   China RoHS

In China, new requirements are made for internal directives similar to those in the EU, commonly called China RoHS, with additional

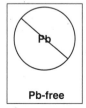

FIGURE 5.2
JEDEC Standard
JESD97 for lead-
free.

Pb-free

FIGURE 5.3
China RoHS
product labeling.

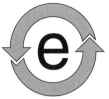

Compliant product   Noncompliant, substance
                         disclosure required

requirements spelled out in "Management Methods of Controlling Pollution by Electronics Information Products - Ministry of Information Industry Order #39." This directive comes with the added burden of product labeling. The labeling is only required of those products being imported to China and not those being exported. Some consider this added burden as protectionism rather than environmentalism. China RoHS has two label identifiers which are required as part of the directive and indicate compliance or noncompliance, as shown in Fig. 5.3.

5.2.5 Electromagnetic Compatibility (EMC) Directive

The EMC directive, called "Electromagnetic Compatibility," EN61131-2 2003, specifies that products sold must not output excessive electromagnetic interference known as *emissions* or that they not be adversely affected by electromagnetic interference known as *immunity*. There is no marking to identify EMC compliance. It requires that screening be done to identify which products are outside the new specifications. This may require the purchase of new software tools to assist the engineer in isolating problem areas in the layout of the printed-circuit board (PCB). Product identified typically requires new replacement components as well as PCB artwork layout changes through the CAD designer.

5.2.6 Waste from Electrical and Electronic Equipment (WEEE)

This WEEE directive, known as 2002/96/EC, is aimed to have designers incorporate methods which lend themselves to easy dismantling of products for recycling at the end of their life cycle. This directive is very different from earlier design for assembly (DfA) initiatives to reduce the number of parts and ease of assembly with less hardware and snap-together designs. Products which comply with the WEEE directive may display a marking such as that shown in Fig. 5.4.

5.3 Global Legislation for Green Products

In the United States, environmental initiatives frequently begin at the state level, and some state legislatures have mirrored the EU directives'

FIGURE 5.4
Product marking
label example
(WEEE).

intent such as California's Electronic Waste Recycling Act (EWRA). Other states such as Minnesota, Maine, Maryland, New Jersey, Tennessee, Vermont, Washington, and Wisconsin have laws similar to RoHS. Massachusetts is close behind with Senate bill 558, "An Act for a Healthy Massachusetts: Safer Alternatives to Toxic Chemicals." With the election cycle in Congress, there is speculation that the United States may see its own version of RoHS legislation in the near future. Other countries are formulating similar laws making RoHS a requirement, to stay competitive in today's global market. Upon investigation, not all products are required to comply with these regulations, and this can help prioritize what products need to be addressed for compliance first. Many companie,s however, especially multinationals, will make the tough decision to eventually make all their products comply with all these directives. Here is a brief listing of other legislations:

- RPCEP (Regulation for Pollution Control of Electronic Products): China, effective July 1, 2006

- JGPSSI (Japan Green Procurement Survey Standardization Initiative): Japan, effective July 1, 2006

- SB20 (Electronic Waste Recycling Act of 2003): California, effective January 1, 2007

- Other countries adopting the EU RoHS directive: Australia, Canada, Korea, Taiwan

5.4 The Global RoHS Implementation Organization

For a multinational company in which there are multiple product groups, manufacturing sites, and logistics groups in multiple countries reporting under a central organization, the organization of the RoHS implementation team is much more complex than for a single-country or single-locality company. It is critical to set up an effective reporting structure and reporting relationships to successfully manage on a global scale.

Two levels of reporting are required, the project progress report to the management at the product department level and a progress report for executive management that is an accumulation of all the product departments to provide a companywide view of implementation progress. It is the responsibility of a global project team to ensure project status reports are accurate and timely each month, to deliver a complete picture of the company's current state of environmental compliance readiness.

The global core team, whose structure is shown in Fig. 5.5, will be comprised of representatives from each product design

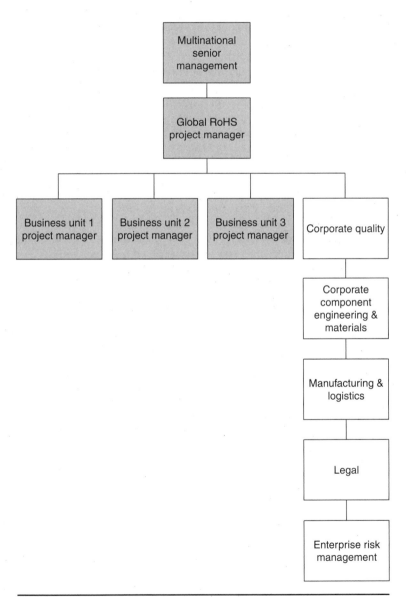

FIGURE **5.5** Structure of global core team.

department, preferably from a project management office, corporate quality, component engineering and database administration, manufacturing, global logistics and warehousing, failure analysis laboratory, legal, and enterprise risk management. Planning should be conducted using accepted best practice program management techniques.

The global project team works initially with executive management to establish the company's policy so the implementation teams have guidelines for planning at the local level. The guidelines should contain information regarding management's interpretation of the directives and how they apply to the company, reaching a consensus on the company's responsibility, identifying product categories or families for which the directives apply and those that do not. A milestone schedule or time line for implementation is required to do proper local resource planning. Each product department should then assign its own project manager to build and lead the local site implementation team.

Quarterly global team meetings are held in a centrally located conference center. Project managers from each of the product departments come to the meetings to provide status, detailed risk assessments, both local and global, and discuss lessons learned with their colleagues. In these meetings, the corporate guidelines are interpreted for potential implementation and pitfalls such as not having the resources to accomplish a task. For example, if management decides to change part numbers of components and parts, it could mean a very large task of documentation and product structure updates. It is vital for the individual project managers to understand the needs of upper management at this phase to advise them on the risks and implications of their policy decisions.

The risk in these global meetings is that miscommunications, misunderstandings, and cultural misinterpretations can provide inaccurate impressions as to the progress being made. There are differences, however slight, in the way people from different countries approach project tasks. If these misinterpretations take hold, the local project manager can spend numerous hours collecting new data to support his or her position and clear up the misconceptions. This can be politically costly to the local manager, since the questioning usually starts at a higher management level and flows downward, so special care should be taken to present information in a cross cultural fashion.

In between these quarterly meetings, accurate reporting is required in the desired formats. The global team is responsible to interpret and summarize the information from each of the product departments and put it into a format which will inform upper management of the organization's progress as a whole. It is also their responsibility to hold local product departments accountable for commitments made throughout the planning process and adhere to the schedule established by the global team and approved by upper management.

5.5   The Local RoHS Implementation Team

At the site level, the local RoHS implementation team is comprised of individuals from several departments. The responsibilities of each department are shown below:

- *Advanced Manufacturing and Test (AMT).* This department is to ensure product design compliance with manufacturability standards, allowing for the lowest-cost product assembly. The AMT group plays a managing role in the redesign process, if so required. The roles include manufacturing process expertise, factory process status quo, in-circuit test (ICT) programming and fixture design, management of the transfer process from design for manufacturing, test analysis, part procurement, factory resources, and finished goods inspection. This group may also have responsibility for the enterprise data management function.

- *Marketing.* This is a key position in both the initial planning and appraisal and later customer communications. Marketing has to indicate which products fit into the categories identified in the directives, and identify which product department is responsible for the designs.

- *Component Engineering.* As products are identified, their bills of materials need to be broken down into individual components to be checked against the company part database and available vendor specifications. Intensive communications will be required with component vendors or manufacturers' representatives to identify the material composition or active sale status of particular parts.

- *Continuing Engineering.* The mission of this department is to handle design and customer issues for products which are already in the marketplace. The department performs initial redesign analysis and analyzes information from Component Engineering to determine the extent of redesign that will be required to comply with the directives and the new compliant manufacturing processes.

- *R&D Resources.* If needed, resources must be diverted from new product design to supplement resource shortfalls or skill set needs.

- *Quality and Customer Satisfaction.* The job of this department is to ensure that the product redesign does not result in a change to the form, fit, or function of the product and that the appropriate process steps are followed. This will ensure that customers will not be adversely affected.

- *Verification and Validation Testing.* These departments ensure that the redesigned product still responds as originally designed and reacts as the typical customer expects.

- *Agency Compliance.* This group manages and ensures that the required compliance oversight organization standards are maintained. Organizations such as Underwriters' Laboratories (UL) ⓤ, Conformité Europenne (CE) CE, and Canadian Standards Association (CSA) Ⓢ previously satisfied the need to be reviewed and at a minimum may need a paperwork update submitted.

An important step in the success of the local team is to create a clear policy document for designers to navigate the new criteria. Issues to address include these: When do the new rules take effect? What is the impact of the change and how does it affect design and manufacturing practices, quality classifications of restricted use and forbidden use components? Special attention is needed to ensure all aspects of the products parts are being aggressively investigated, there can be a tendency for less exciting parts such as screws, nuts, and bolts to be ignored unintentionally which will become a potential problem later when compliance assessments are reviewed.

5.6   The Product Industrialization Team

Today, companies are working in virtual design organizations with increased interdivisional projects that are geographically distributed. A global virtual team is comprised of members with distributed expertise and skills that span the boundaries of time, geography, nationality, and culture. If the production capacity is located in Europe or Asia, the manufacturing, process, materials labs skills, and logistics expertise will be located there as well. The product industrialization team is responsible to plan and implement new products at the factory level and interfaces with the product department industrialization leaders in other business units. In addition, the team leader interfaces with the following manufacturing disciplines, as shown in Fig. 5.6. Her or his focus is to develop and implement compliance with directives. The product industrialization team should be comprised of the following skill sets.

1. Process engineering for manufacturing process expertise

2. Failure analysis and materials to investigate and test, reliability of new materials and processes and to assess future quality and reliability specifications and issues.

3. Procurement/vendor management to establish and maintain agreements with outside suppliers of components and assemblies.

4. Purchasing to manage ongoing part procurement based on production requirements.

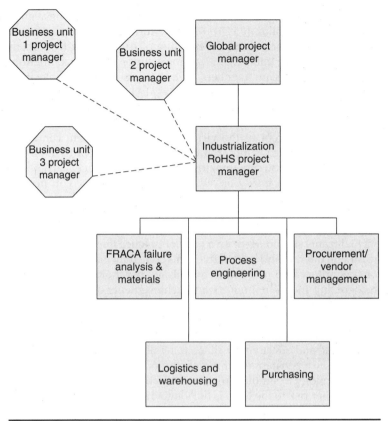

FIGURE 5.6   Local Project Team Structure.

5. Logistics and warehousing to determine plans to integrate an effective inventory transition from non-RoHS- to RoHS-compliant products. Will components be segregated or mingled in the same stock location with a unique identifier? How do warehouse personnel identify compliant products versus noncompliant products? Will sales of both products be allowed? Can overstocks be sold in countries in which the directive does not apply or will they be scrapped? If scrapped, at whose expense?

If there is a need for overseas design, site teams will be identified to complement the local design organization skill sets.

Advancements in communications and computer technology have made it possible to interact on a global scale. Tools such as net meetings and video conferencing allow teams to actively trade information, without ever having a face-to-face meeting. These new technologies give rise to the risk of poor-quality communications and information sharing. Virtual teams do not develop the cohesiveness experienced by collocated groups. Few forms of informal relationships develop which aid in the building of trust among team members. Virtual team interactions can foster unproductive competition for status among team members. Due to the increased complexity of the team dynamics, emphasis should be placed on the importance of focusing on team development, ensuring team members have a good fit between personal and project intent.

With the concept of the virtual or distributed team comes the issue of trust, which is important in all relationships but with a special focus here. If you sign up with a resource or resource(s) in another geographic region inside or outside the organization, you need to develop trust that an individual or group will act in the best interests of the project. If team members can meet face to face early on, that will greatly increase success in building trust within the team. The virtual teams should agree on a code of practice that sets out how the team will interact and be effective. This code may include a fixed time or time interval to respond to e-mails, or psychological support by recognizing individual efforts and achievements on the part of the team. This is consistent with the ground rules for internal teams and should be supported in the virtual team. It is important to be aware of cultural practices that may influence the code of practice so that it is valuable and effective for all team members. Teams should exploit diversity, use constant communication to hold a team together, and use technology to simulate reality as much as possible.

Another risk to the team is the longevity of the members. A member or members of the team may leave anytime for any reason. A plan is needed to deal with this adverse potential. In addition, vacation and holiday practices, especially in Europe, should be understood. A supplier may be located in an American Indian community where traditional spiritual practices are followed which do not synchronize with typical product planning cycles.

5.7 Communications among Global Teams

When working on global teams, it is not always possible to meet frequently due to scheduling and travel costs. The level of cooperation with other units will go up considerably after face-to-face exchanges and sharing. If this is not possible, team members can participate through the use of tools such as NetMeeting, WebEx, or other video conference packages. This may mean getting up very early to participate depending on the time zone in which the meeting is being conducted.

The global team leader has a challenging position: it is her or his responsibility to take in all the concerns and issues for local sites, summarize them, and discuss the issues with top management. This will result in implementation policies, goals, and timetables for the product departments. Upper management must demonstrate its commitment to the goals by communicating through the various means mentioned above to share their vision and address questions and concerns of the local leadership, once that leadership has absorbed the tasks at hand and their impact on the project portfolio.

For the local project manager, there is a challenge in meeting with remote factory team members to stay current with progress toward the training, equipping, and conversion or setup of RoHS-compliant assembly processes and track process change needs. It would be very risky to assume that these activities will take care of themselves or are the responsibilities of someone else. When a factory is late in implementing RoHS, it would be advisable to have identified and qualified a contract manufacturer to bridge schedule gaps and deliver compliant products to customers on time. The larger contract manufacturers have been aggressively implementing RoHS processes and may be further along than the captive operations.

5.8 Understand Which Products in Your Portfolio to Redesign

Many companies have gone through at least one or several acquisitions. Products get moved around and remixed to fit the new product portfolio of the parent company. A product in a company's portfolio may be a design responsibility of one group but is actually marketed by another group, or is even a third-party label. A first step is to compile a list of what the team is responsible for from a design and/or marketing perspective. The team to do this should include members from design, marketing, and project management. If third-party suppliers are involved, then the procurement organization should also be part of the team. In a multinational company, communications with counterparts in other product groups or subsidiaries are needed to reach definitive agreements with regard to who will take the lead on shared products. The decision on this is normally based on available skill sets and then resource availability. The results of this activity would be a listing of information such as the following:

1. End item product part numbers
2. Major subassembly part numbers
3. Product design responsibility owner
4. Marketing responsibility owner
5. Product sales code: active, restricted, service only, end of life
6. Annual sales quantity
7. Product type and family
8. Responsible product manager
9. Supplier, if brand label or third-party vendor

Some top-level issues should also be discussed because they affect practices throughout such as these:

- Will individual part numbers for piece parts be changed if the parts supplier provides a RoHS-compliant direct replacement part?
- Will assembly or model numbers change if the product is redesigned to be compliant but there is no change in form, fit, or function?
- Will the product be marked for customers to identify a compliant product or will it be simply marked in some fashion to assist internal logistics personnel?
- What type of customer communication is there and who is the responsible group to centrally handle enquiries?

Part number changes need careful consideration since it will require significant effort updating part databases, bills of material, purchasing records, warehouse space, etc. Model number changes will affect product databases, logistics and warehousing, sales and marketing brochures, and country sales teams. A typical RoHS decision to redesign process flow is shown in Fig. 5.7.

5.9 The Project Directory for Local Control

Availability of information such as directives, policies, project status, and day-to-day operations is critical to success. A project should be created and be well known to everybody on the team and allow the appropriate level of permission. It should contain a project status worksheet for each phase being worked by the team. Categories could be like the following:

- Master list of identified product requiring redesign
- Team contact information list
- Product redesign priority ranking
- High-risk components' status
- PCB design investigations
- PCB design status, in-process and completed
- Status of nonactive components: cable and harnesses, mechanical assemblies, plastic parts and hardware
- Brand label modules, cables and harnesses
- PCB through-hole annual ring-shape footprints
- ICT fixtures and test program updates

A typical local project status and history log is shown in Fig. 5.8.

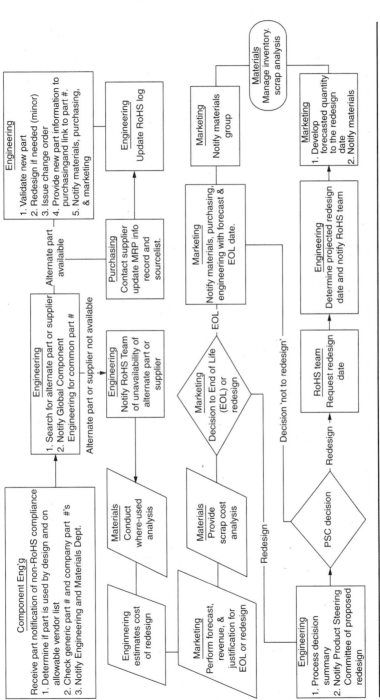

FIGURE 5.7 RoHS decision to redesign process flow.

Last update: _(Enter date)_   RoHS Product Redesign Status Chart

List maintained by:

RoHS priority	Product (Model) part number	Description	Redesign phase	Tin whisker parts evaluation completed Y/N	Status comments	ICT status	Project	Existing PCB#	New PCB#
2		Printed circuit assembly XX	Unit test	Y					
1		Printed circuit assembly XX	Investigation	N					
2		Printed circuit assembly XX	Unit test	Y					
1		Printed circuit assembly XX	Not started	N					
1		Printed circuit assembly XX	Not started	N					
2		Printed circuit assembly XX	Prototype build	Y					
2		Printed circuit assembly XX	Prototype build	Y					
3		Printed circuit assembly XX	Investigation	N					
1		Printed circuit assembly XX	Investigation	N					
3		Printed circuit assembly XX	ECAD Layout	N					
2		Printed circuit assembly XX	Investigation	N					
3		Printed circuit assembly XX	Design Hold	N					
2		Printed circuit assembly XX	Unit test	Y					
2		Printed circuit assembly XX	ECAD Layout	Y					
2		Printed circuit assembly XX	ECAD Layout	Y					
2		Printed circuit assembly XX	ECAD Layout	Y					

FIGURE 5.8 Project product redesign status chart.

This directory method should be expanded to include tracking of effected products, at-risk components, cables and harness, OEM manufacturers, etc., wherever there is risk of exposure and a need for tracking daily or weekly. A journal of the history of the project is also helpful to document the evolution of the project and design decisions.

Product redesign is a difficult challenge, and it is very improbable to hold to original plans and expenditures, since the requirements invariably change over time, as new directives come into force, component vendors postpone production of or make obsolete significant parts, or reliability data for new parts fail established reliability targets. Other useful techniques include building a library of white papers and technical articles on design, manufacturing process, and reliability for compliance issues.

5.10   Components Obsolescence Resulting from Green Conversion

The RoHS implementation by component manufacturers has necessitated early obsolescence based on market demand versus cost of conversion. If a component becomes obsolete, it may result in PCB layout changes and incompatible board support packages (BSPs) for active components, materials, and process risks such as tin whiskers. Below is a sample of a website that can be used to assist in identifying potential component issues.

Components Obsolescence Website: PNC Alert

http://www.pcnalert.com/Marketing/

Other websites are component data mining resources available to designers as well as component news web suites, shown in Fig. 5.9.

5.11   Quality and Reliability

Many component vendors have replaced traditional tin/lead solder alloys with pure or high-tin-content alloys. This presents designers with a potential long-term quality issue of tin whiskers, which are tiny tin filaments that grow over time with the potential of causing electrical shorts in equipment. If products have a long service life, or are used in harsh environments with temperature and humidity extremes, or are used for critical operations, tin whiskers are a serious problem which cannot be ignored. JEDEC Standard 201 is used as the testing standard to qualify components. It is advisable to obtain a copy of this standard for use in preparing an acceptance procedure for electronic components. Figure 5.10 lists key industry specifications for lead-free electronics.

Components search–Datasheets

www.alldatasheet.com
Asia Global Sources Electronic Components
    Asia's largest database of components and suppliers
Chip center
Connector specifier
Datasheet locator
Faradnet
IC Master
Part miner-free trade zone
Questlink
Semiconductor datasheets on the web

■ **News Sites**
EE times online
Electronic news
Electronics supply & manufacturing
Electronics weekly
EDN
Electronic business
Electronic design online
Electronic product magazine
Electronics web
PlanetEE
Semiconductor international

FIGURE 5.9   Information resources.

Component engineering and quality will be required to develop go/no-go criteria with respect to electronic component finishes. It is advisable to separate them into three categories:

1. Rejected tin finishes – not to be used under any conditions

2. Tin finishes which will be acceptable provided the component manufacturer provides test results showing the lead finish performed in an acceptable fashion as specified in an industry accepted test specification such as JESD22A121 or 201

3. Acceptable tin finishes without supporting test data

Fine pitch (active) components are of particular concern, and acceptability standards must be determined prior to launching product designs.

This acceptability decision is commonly made by a central materials lab, and the information is stored in the component engineering part database for engineers to refer to when selecting parts. Each product group needs to set up a review process for components

Document Identifier	Description
IPC/JEDEC-020B	Moisture/reflow sensitivity classification for non-hermetic solid state surface mount devices, july 2004
IPC-A-610D	Acceptability of electronic assemblies
JEDEC Standard JESD22A121	Environmental acceptance requirements for tin whisker susceptability of tin and tin alloy surface finishes, march 2006
iNEMI	Recommendations on lead-free finishes for components used in high-reliability products (Version 4, updated December 1, 2006), INEMI tin whisker user group

FIGURE **5.10**   Key industry specifications for lead-free electronics.

because in many cases, the materials laboratory has not had the opportunity to review every part. This mitigates the risk of selecting a forbidden lead finish in the design, thus rendering the product non-compliant. As an individual part status is determined, the data should be submitted to the central materials lab for approval and archived to the component database.

A review process is needed to resolve problems such as a component being labeled as restricted in error, when there are data from the manufacturer to the contrary. Experts need to be identified to review the part specification data against the established standard for acceptability. If the component is found to be acceptable, a decision must be made to alter the status, document the decision, and present it to a designated member of senior management, preferably the director of quality for approval and submittal to the materials laboratory.

5.12   Lead-Free Assembly Process

It is necessary to understand the changes to the process, equipment, and training when converting to lead-free soldering. A reasonable method is to join a local industry group or research consortium and collaborate with others who have the same needs to conduct process experimentation and testing. Teaming up with university academia, state environmental resources, and industry experts cultivates an open forum for brainstorming and information sharing sessions that are of great benefit to advancing your understanding of the characteristics of the new materials.

Design engineers are concerned that selected components be able to withstand the elevated-temperature profiles used in the lead-free solder reflow ovens and wave solder machines. Additionally, they need to make accommodations for component spacing and annual ring size for bottom side components which will go over the wave solder machine.

Assembly equipment needs to be reviewed to meet the higher temperature requirements. The solder reflow ovens of model year 2000 and later can be profiled if the PCBs are not densely populated or greater than six layer laminates. Otherwise it is advised that new ovens with additional heating zones and elevated-temperature capabilities be purchased. The use of a nitrogen atmosphere is a matter of process expertise and quality preference.

The wave solder machines require either replacement or retrofitted solder pots and nozzles. An issue is the use of older solder pots which are not compatible with lead-free solder alloys and will degrade over time. This degradation results in leaking pots and can affect the paddles of the solder pumps which will make calibrating the wave dynamics nearly impossible. Retrofit is commonly an acceptable option depending on the machine supplier. A retrofit will consist of a new solder pot, nozzles, and solder pumps.

Hand soldering personnel will require new training because the skills in tip temperature selection and dwell times are slightly different with lead-free soldering, and operators will need to adjust their skills to the new material for optimal results.

Inspection personnel are going to need to recalibrate their methods toward slightly different definitions of a soldering defect. Lead-free solder joints have a different appearance. Training will need to be conducted to comply with the latest revision of IPC-A-610D, "Acceptability of Electronic Assemblies."

Moisture control methods for PCBs need to be reviewed due to the new process temperatures required. The International Electronics Manufacturing Initiative (iNEMI) has a standard called JSTD-020B, "Moisture/Reflow Sensitivity Classification for Surface Mount Devices."

5.13 Communicating with and Qualifying Local and Overseas Manufacturing Facilities

In many global design center organizations, there is the advanced manufacturing team which acts as a liaison between the design site and the overseas manufacturing facilities. Much of the staff has significant practical hands-on experience which was gained before products were moved to lower-cost countries for production. It would be beneficial to consider developing a local partnership with a small contract manufacturer. This will provide both local quick turn assembly capability for use during product design and development and hands-on process experience for local engineers in the lead-free process. This may require some additional investment in equipment and resources and many hours spent implementing the process at the partner facilities.

Process testing is a valuable tool in this effort if it can be set up locally. It provides an opportunity to test the materials being specified by the corporate materials lab as well as to keep the skills of the manufacturing engineers current in current materials and production processes. This will help when interacting with global colleagues and overseas contract manufacturers. By having a local contract manufacturer, testing can be performed on solder pastes, wave, and hand soldering materials. Material characteristics can be also determined such as solder paste workability, dispensing and shelf life, product odor and its possible effect on operators, print characteristics of paste height and volume, stencil aperture clogging over time, and frequency of print cycles before cleaning. Similar testing could be performed for wave soldering and hand soldering based on the process specific characteristics.

Qualification of lead-free RoHS-compliant processes for local and overseas manufacturers for surface mount technology (SMT), wave solder, hand soldering, and rework requires a detailed approach. The testing should be done in stages. The first process development should be SMT. New solder pastes should be tested to ensure they produce high-quality soldering results in a oxygen or non-nitrogen atmosphere, ensuring no systematic changes as a result of the new material. The paste that is chosen must also be compatible with conformal coating processes.

The second test stage, wave solder, will require a less detailed test. The solder alloy of choice will be dictated by the corporate materials lab to ensure companywide standardization. Small lots should be tested in small runs with the new material to ensure the results meet current quality standards. The company must be committed to upgrading the existing wave solder equipment to comply with the lead-free alloy for material compatibility and wave dynamics.

Finally, rework and hand soldering should be tested. This testing may be very limited in scope. The only new requirements for these operations will be the use of new solder heating tips and the new solder alloy chosen by the corporate materials laboratory. Typically, the same alloy for wave solder will be used for hand soldering and rework. It may not be necessary to change the no-clean flux for hand soldering and rework. The only potential change for this type of soldering will be an increase in tip temperature to the 700 and 800°F tips versus the 600 and 700°F tips used today. However, this requires experiments as some companies have found the dwell time to be as much of a factor as tip temperature. The company must be committed to operator training on the new lead-free standards and requirements.

5.14   Product Compliance Information Decisions

Compliance could be indicated by properly marking the product. Product marking is mostly dependent on available space on the product and packaging. It may also be a marketing decision dependent on the view of how product marketing will influence the company's competitive advantage versus the cost to operations and sales teams, including the choice of not physically marking the product with some form of industry acceptable identifier.

Markings may also be a requirement of the specific directive. For example, the China RoHS directive requires visible markings on the product which specify if the product is compliant.

At the same time, enquiries from customers as to the status of the products need to be addressed. At the corporate level, there should be a consistent, unified response along with a record of all customer enquiries from all over the world. Requests may be tallied by reference of top-level model numbers or down to piece part data requests. Some manufacturers, for example, maintain a zero tolerance for parts using hexavalent chromium. Customers with such policies may want to determine the contents of the product they are buying, down to the lowest part level. A central database located on a customer service website could be available to customers to enter a model number, check a date code or revision level, and obtain the RoHS compliance implementation data.

5.15   Brand Label Management

Products which are designed and manufactured by another company to the company specifications and then labeled with the company logo are *brand label products*, and the company is called a brand label company. In this case, the is a need to identify the people in the organization involved in the management of those products: marketing defines the requirements; program management along with purchasing may manage the relationship with the brand label, including production forecasts, budgets, and the brand label account manager on the other end. Accomplishing a design change of this environment relies on the business relationship, the sales success of the current product, inventory levels of finished goods in the warehouses, piece part inventory counts purchased by the brand label company on a risk basis, and anticipation based on the supplied production and sales forecast. A plan should be agreed upon with the OEM for arranging for conversion to green designs and processes. A phased plan should be rolled out in the following manner:

- When does the redesign need to be completed and who is paying for the effort?
- What is the date of availability to the sales force?
- Will both compliant and noncompliant products be offered at the same time?
- Will there be scrap of finished goods or piece part inventory, and who pays for it?

- Does the OEM process meet the company's quality requirements, such as the use of high-temperature laminates and plastics components capable of surviving the elevated thermal profiles of lead-free soldering?
- What solder type and alloy are specified for your product?

A reciprocal arrangement between the company and the OEM agreeing to assist each other in successfully meeting the goals is required. It may include frequent site visits to the brand label manufacturer to ensure that all parties are on the same page and that everybody is delivering on the commitments made during planning. It is advisable to work with purchasing, get written commitments, and not settle for verbal assurances. There is no substitute for regular communication. The goal is to not interrupt product availability and to avoid any negative customer impacts.

5.16 Data and Documentation Requirements

Multinational efforts require large amounts of data collection and management. Representation from these groups may be part of the global core team. Ideally, there are common tool sets across organizations; otherwise the global team leader needs to plan the data management from the local level and compile it for use internationally. The database should include updates to product structures, part numbers, classifications and specifications, approved suppliers, design drawings, components, etc. Since the global team services several internal organizations, plans need to have realistic assessments of cycle times for investigations and decisions on database updates. If redesign and current production schedules do not match, it is the local team's responsibility to assess the risks involved and put into action a plan which bridges the gaps and manages the design's evolution.

Problems could multiply if the company has not been diligent in keeping older documentation up to date with the latest tools used for product data management (PDM). Examples would be drawings still in paper format stored in drawers and cabinets, and several may be missing altogether. There are many options to resolve these problems: convert paper drawings into electronic documents, relying on 'as-built' bills of materials from the company's operations or suppliers, or update every document to reflect its RoHS compliance status. There is a need to establish clear ground rules on how to handle each problem category so as to avoid confusion and moral issues.

5.17 Risk Management

Managing risk is about managing opportunity. There is no possibility of eliminating all risks, but the impact and rewards for each step of the product plan should be recognized. In the case of RoHS, it involves creating competitive advantage in the marketplace. Classic risk management texts look at the five W's and an H, when analyzing the objectives of any plan.

- *Who?* Which departments have product design and marketing responsibility? Who are the customers and what are their needs?

- *Why?* Based on strategic needs, which products should be targeted for compliance, in what priority? What happens if you do not have a compliant product to sell? Are there customers who perform their system certifications at their site and who do not want compliance products?

- *What:* What types of products and sources are there—OEM, outsourced to contract manufacturers, or brand label?

- *Where?* What are the target dates with which products should comply? And what happens at different locations and where? What happens to products on the shelves of distributors and warehouses?

- *When?* Is the organization ready to declare compliance? Can it monitor performance?

- *How?* What resources are required to accomplish project goals? How will you measure output and success?

Once these questions are asked, decisions can be made. It might be that it is decided to do nothing because the risks do not warrant the investment in RoHS conversion. The motto should be: "We are prepared, we are aware, and we have considered and accepted the causes and effects."

5.18 Conclusion

This chapter is the product of the author's experience in managing the implementation of green design practices resulting from government regulations and directives. The ideas and techniques presented can assist in successfully planning and executing similar green projects in design and manufacturing operations. It involves virtually all segments of the organization; it is a great networking opportunity for the project manager, especially if it is for a large multinational with sites distributed throughout the world. Teamwork, resource skills, clear direction, and support of management are required throughout the entire process. Do not expect this to be a short-duration project; it is a continuously evolving process of discovery and replanning. It is critical to satisfying customer needs; and as difficult as it is for your organization, it is just as difficult for many of your customers. Having a ready solution increases your organization's competitive advantage both to the external customer and to the internal ones.

References

Ball, D., et al. *International Business, The Challenge of Global Competition,* 10th ed. New York: McGraw-Hill Irwin, 2006.

Mackenzie, K., "Business Must Reduce, Re-Use, Recycle," *Financial Times,* September 19, 2006.

Maitland, A. "'Virtual Teams' Endeavor to Build Trust." *Financial Times,* September 8, 2004.

Majchrzak, A., A. Malhotra, J. Stamps, and J. Lipnack. "Can Absence Make the Team Grow Stronger." *Harvard Business Review,* May 1, 2004.

Industry Standards and Websites

"Environmental Acceptance Requirements for Tin Whisker Susceptibility of Tin and Tin Alloy Surface Finishes," JEDEC Standard JESD22A121, March 2006.

http://leadfree.ipc.org/RoHS_3-2-2.asp

http://www.bradyeurope.com/web/SiteBuilder/BradyEurEN-SEO.nsf/FLV/Labels+-+Chinese+Rohs

http://www.jedec.org/download/search/JESD97.pdf

http://www.weeerohs.bradyeurope.com/web/SiteBuilder/Europe72ISBv1r0.nsf/FLV/weee#weelie

iNEMI, Tin Whisker User Group. "Recommendations on Lead-Free Finishes for Components Used in High-Reliability Products," Version 4, updated December 1, 2006.

IPC-A-610D, "Acceptability of Electronic Assemblies."

IPC/JEDEC-020B, "Moisture/Reflow Sensitivity Classification for Non-Hermetic Solid State Surface Mount Devices," July 2004.

JESD22-A111, "Evaluation Procedure for Determining Capability to Bottom Side Board Attach by Full Body Solder Immersion of Small Surface Mount Solid State Devices," May 2004.

www.jedec.org, JESD201, "Environmental Acceptance Requirements for Tin Whisker Susceptibility of Tin and Tin Alloy Surface Finishes," JEDEC Solid State Technology Association, revised March 2006.

CHAPTER 6

Successful Conversion to Lead-Free Assembly

Robert Farrell and Scott Mazur

Benchmark Electronics, Hudson, New Hampshire

6.1 Introduction

The driving force for the conversion to lead-free manufacturing and recycling is two directives mandated by the European Union (EU) in July 2006. The first is the Restriction of Hazardous Substances (RoHS) which restricts the use of lead (Pb), mercury (Hg), cadmium (Cd), hexavalent chromium (Cr6), polybrominated biphenyl ethers (PBEs), and polybrominated diphenyl ether (PBDE). The restriction of lead has the greatest impact because most electronic products are assembled with lead-based solders. The other substances are also important because they can be found in printed-circuit boards (PCBs), components, cables, etc., but their impact is not as great as that of lead. The second directive passed by the EU is the Waste from Electrical and Electronic Equipment (WEEE) directive which mandates recycling. It does not directly affect manufacturing but can influence design in the selection of components and design of the product to facilitate recycling. Similar directives have been passed by other countries including China, Japan, and South Korea.

The conversion to RoHS compliance (lead-free) is one of the greatest changes to the electronics assembly in over 20 years and exceeds the impact of the more recent conversions of through-hole to surface mount technology (SMT) and high-pin-count lead frame components (plastic quad flat pack or PQFP208) to ball grid array (BGA) packages. This is true because lead-free affects the entire process including component acquisition, assembly, test, rework, field service repair, and documentation. The transition is gradual which means tin/lead and lead-free assemblies will likely have to be manufactured in parallel at each manufacturing site, and extensive process controls are required to ensure the integrity of the lead-free and tin/lead products within the same manufacturing site.

The RoHS directive includes a number of product and component exemptions; if a company is included in the exemption, the conversion could be delayed. However, it may become more difficult to assemble tin/lead products in the future because of the inability to obtain tin/lead components (especially BGAs) in conjunction with increasing global environmental restrictions and the potential that customers may require lead-free products. Lead-free BGAs are not drop-in compatible with the tin/lead process, and companies may be asked to become green by their customers. Preparations for conversion are recommended for all companies including those whose products may be exempt.

A successful conversion to lead-free production requires good preparation and planning prior to the start of volume production. This chapter summarizes the steps taken by a major electronic manufacturing service (EMS) company to convert to lead-free. The information in this chapter is for use as general guidelines only, and they are subject to change based on continuous lead-free improvements. Interpretation, deployment, and modification of these guidelines for lead-free assembly conversion should be altered or modified depending on each specific requirement or potential use in a product.

6.1.1   Planning for the Conversion—An Overview

A tremendous amount of lead-free research and development has taken place since the early 1990s, and a central focus of the conversion was to build on this work. This minimized risks in terms of assembling a reliable product, simplified the conversion, reduced the costs, and ensured consistency with emerging industry standards and practices. The conversion process included the selection of cost-effective test vehicles for initial validation, then a larger, more complex test vehicle, and participation in a lead-free consortium. The latter was the New England Lead-Free Electronics Consortium managed by the Toxic Use Reduction Institute (TURI) (http://www.turi.org/industry/research/supply_chain_program/electronics) which involved academia, government, and industry. The applicable government agency was the U.S. Environmental Protection Agency (EPA), and industry members included component, stencil, and PCB suppliers, as well as high-reliability original equipment manufacturers (OEMs). The latter are typically customers of an EMS provider, and the consortium provided valuable exposure to the complete supply chain. This is important because a successful lead-free conversion requires coherence and teamwork throughout the entire supply chain. An EMS company, such as Benchmark Electronics has a primary focus of ensuring the reliability of the lead-free solder joints, the components, and the PCBs. They must be selected and specified to withstand the higher temperatures associated with lead-free manufacturing.

6.1.2 Selection of the Lead-Free Alloy

The first decision that must be made is the selection of the lead-free alloy to replace the tin/lead eutectic alloy. The National Center for Manufacturing Sciences (NCMS) initiated a project in the mid-1990s to determine feasible replacements. NCMS research indicated that a Sn3.5Ag alloy had been the choice for many manufacturers and had mechanical properties equal to or better than those of the legacy tin/lead eutectic solders.

NCMS started with approximately 200 lead-free alloys and identified 32 that met the following criteria:

- Did not contain toxic elements
- Did not fall outside preestablished criteria for manufacturing quality, economics, or reliability
- Met certain metallurgical criteria

A reliability test vehicle (RTV), which incorporated a variety of SMT components, was thermally cycled between –55 and 160°C and narrowed the number of alloys to seven. The seven lead-free alloys outperformed the reliability of the tin/lead baselines on BGAs but underperformed on the 20-pin leadless ceramic chip carriers. Overall, the seven alloys were considered as viable tin/lead replacements. Two of the alloys included the SnAgCu (SAC) ternary compound with a variation in the silver content (3.0 and 4.0 percent). Both of the SAC alloys performed equivalently which suggested a relative insensitivity to variations in silver.

Further studies indicated the addition of copper to the tin/silver alloy yielded the following benefits:

- Reduced the melting point by approximately 3°C
- Slowed the copper dissolution in the circuit board
- Improved wettability, creep, and thermal fatigue as compared with the tin/silver

Studies also indicated a tin/silver/bismuth (SnAgBi) alloy provided additional improved reliability as compared with the SAC alloys, but there was a dramatic drop in reliability if a lead-based material (board or component surface finish) contaminated the alloy. The alloy was not considered a viable lead-free alternative at this time due to the uncertainty regarding 100 percent lead-free compliance on the part of the component and board vendors. In addition, bismuth is a by-product of lead mining.

Following the NCMS initiative, the International Electronics Manufacturing Institute (iNEMI) conducted additional development work and recommended a SAC alloy comprised of 95.5 percent Sn, 3.9 percent Cu, and 0.6 percent copper as the standard lead-free replacement for tin/lead.

The Solder Product Value Council (SPVC), which is an Institute of Printed Circuit (IPC) committee, continued with additional testing and recommended a SAC comprised of 96.5 percent Sn, 3.0 percent Ag, and 0.5 percent copper which is commonly referred to as SAC305, where the "30" means 3.0 percent Ag and the "5" means 0.5 percent copper. The NCMS and SPVC work both confirmed the SAC305 and SAC396 (iNEMI) alloys were equivalent, but the SAC305 was selected by the SPVC partially because of its prevalence, particularly in Asia.

Based on the above work, Benchmark Electronics elected to standardize on SAC305 for all lead-free assembly including SMT, through-hole, touch-up, and rework. This decision greatly simplified the work required by eliminating the need to evaluate different alloys. Equally important, it ensured consistency with the entire industry especially with regard to plastic BGAs. Most lead-free plastic BGA manufacturers are using SAC305 spheres which means they melt in the lead-free assembly process at the same temperature as the lead-free solder. A different lead-free alloy would have melted at a different temperature as compared with the lead-free spheres, which might have introduced quality and reliability issues. At the time of this writing, some BGA manufacturers are using SAC105 spheres which have less silver as compared with SAC305 spheres and there is movement toward replacing the SAC305 in wave solder, selective solder, and rework solder fountains with another lead-free alloy primarily due to the tendency of the SAC305 to dissolve copper pads, especially during rework. Solutions to the copper dissolution issue during rework are discussed in Sec. 6.7, and it is recommended that the reader remain abreast of new developments and consider adopting a new alloy if the industry standardizes on it. The replacement alloy with the largest market share following SAC305 is the Sn100C offered by Nihon Superior. It is comprised primarily of tin with smaller amounts of nickel and germanium and melts at 227°C which is approximately 10°C higher than the melting point of SAC305.

6.1.3   Impact of SAC305 on the Assembly Process

The SAC305 alloy melts between 217°C and 221°C as compared with the tin/lead alloy which melts at 183°C. This means the temperatures during lead-free reflow will be approximately 30°C higher when compared with tin/lead reflow. The molten SAC305 solder is also used in wave solders, selective solders, and solder fountains which are required to solder and rework through-hole components. The temperature of the molten solder in these platforms will also be up to 30°C degrees higher as compared with the equivalent tin/lead platforms.

A true lead-free solution requires PCBs and components to be RoHS-compliant (lead-free) and compatible with the higher processing temperatures associated with lead-free. Specifically, a reflow temperature rating of 260°C is required as compared with the typical temperature rating of 230°C associated with tin/lead. The following components are affected:

- Bare (raw) boards
- SMT components
- Through-hole components
- Press-fit components
- Labels (bare codes, assembly number, etc.)
- SMT adhesive (components and wire)

Press-fit components are included on the list even though they are not reflowed. These components may be on the board when it is reworked, and the higher temperature rating ensures they will not be damaged by the heat or rework. Tie wraps must be RoHS-compliant, but the temperature rating is not critical because they can easily be removed prior to rework and they can be installed after all assembly has been completed.

Shipping material such as electrostatic discharge (ESD) bags, wooden crates, and metal wraps are not required to be RoHS-compliant because the RoHS directive only covers material that is part of the product at the time of discard.

6.1.4 Objectives of the Conversion

The objectives of the lead-free conversion are to provide a lead-free alternative assembly method for customers that meets the following criteria:

- Reliability of end products meets or exceeds that of legacy tin/lead.
- The method is cost-effective.
- No-clean (NC) and organic acid flux formulations are used.
- The method is qualified for most PCB finishes such as organic solderability preserve (OSP), electroless nickel immersion gold (ENIG), immersion silver (ImAg) and Lead-Free Hot Air Solder Leveling (HASL).
- The method is compatible with air atmospheres in the reflow oven.

An OEM may elect to qualify the process for NC only if it is known with certainty that all products can be assembled with NC chemistry. However, an EMS company typically must support both NC and organic acid formulations. An NC flux is one that is not harmful to the product after assembly, and the residue does not need to be removed from the PCB. An organic acid flux residue is harmful to the PCB and must be removed after assembly with an aqueous wash. The four PCB surface finishes mentioned above are expected to be the most common in lead-free assembly, and the requirement for compatibility with an air atmosphere in the reflow oven provides a more cost-effective solution. Most reflow ovens can utilize an air or nitrogen atmosphere with the latter minimizing oxidation during the reflow process. An additional constraint imposed on the conversion is compliance with class 3 of the IPC-610 standard: "Acceptability of Electronic Assemblies" visual inspection criteria. Class 3 is the highest level of visual inspection quality compared with classes 2 and 1. A typical class 3 product may be a computer server whereas a typical class 1 product may be a low-end electronic game. This ensures automatic compliance with all levels, even if the EMS company typically builds product to all three classes.

6.1.5   Product Mix

The lead-free solution must be compatible with the following product mix:

- Double-sided SMT (two reflow passes)
- Area array devices including plastic ball grid arrays (PBGAs), ceramic ball grid arrays (CBGAs), and direct chips attach (DCA) components
- Fine-pitch devices (0.5 mm)
- Mixed technology of SMT and through-hole components on the same board
- Through-hole press-fit connectors
- Board sizes up to 18 to 20 in (457.2 mm to 508 mm)
- Board thickness up to 0.125 in (3.175 mm)
- Board surface finishes: OSP, ImAg, ENIG, and Lead-Free Hot Air Solder Leveling (HASL).
- IPC-A-610, Revision D, "Acceptability of Electronic Assemblies," class 3. Revision D is the latest revision as of this writing and includes inspection criteria for lead-free.

It is important to include key aspects of the product mix in the development activity. For example, solder pastes may perform well on some board surface finishes but not as well on others. All four surface finishes should be included in the development to ensure that the solder paste and processing selected will be acceptable for all. An RTV board that is thick in order to replicate actual PCBs is very important in selecting a qualified flux for wave solder, since thin (and thus acting as lower heat sink) PCBs are easier to solder.

6.1.6 Lead-Free and Flux Chemistry Requirements

The following information was used in selecting the lead-free and flux chemistry to be evaluated. It provides more details than does Sec. 6.1.5, and compliance with these criteria ensures that a robust lead-free solution will be developed that is effective for all products assembled in multiple locations globally with varying environmental regulations.

i. Technology mix:

1. 18- by 20-in Boards
2. Up to 0.150 in thick
3. Up to 36 layers
4. Up to 2.5 oz copper foil
5. IPC 610, Revision D, class 3 criteria
6. BGA/CBGA
7. μ-BGA (micro-BGA)
8. 50-mil Pitch SMT
9. Fine-pitch SMT (primarily 0.5 mm with some 0.4 mm)
10. Through-hole connectors
11. Through-hole power modules (high heat sink) with 0.055-in-diameter leads
12. Components such as 1206's, 0805's, 0603's, 0402's, 0201's
13. Board surface finishes: OSP, ENIG, and ImAg
14. Double-sided SMT and wave solder
15. Wave solder boards to be primarily selective solder with heavily shielded high-heat-sink pallets
16. Rework process of SMT and through-hole components

ii. Solder paste test criteria and standards

1. Bellcore-compliant (NC)

1.1. Passes Bellcore cleanliness testing after aqueous wash for organic acid fluxes

1.2. Testing by a recognized third party is preferred

2. Passes J-Std-004. The definitions of the following tests are given in the standard as well as the pass/fail criteria. A flux must pass all the J-Std-004 tests to be considered by Benchmark Electronics. Fluxes that fail any of these tests are viewed as having unacceptable risks of electromechanical failures in the product during its life cycle in the field.

2.1. Copper mirror corrosion

2.2. Silver chromate

    2.3. Fluorides by spot test

    2.4. Surface insulation resistance (SIR). Minimum allowable criterion is 1×10^8, although Benchmark Electronics obtains a range of readings in the order of 10E8 to 10^{11}

    2.5. Testing by a recognized third party is preferred.

3. Passes J-Std-005.

    3.1. Slump test

    3.2. Solder ball test

    3.3. Testing by a recognized third party is preferred.

4. Good printability (down to 0.4-mm pitch)

5. A 4-h stencil life (minimum), 8-h stencil life (preferred)

6. A 20-min idle time (minimum), 40 to 60 min preferred

7. Hydrophobic

8. Robust in the ranges of 60 to 85°F and 30 percent RH to 70 percent RH.

9. Ease of stencil cleaning and misprinted board cleaning (recommended equipment and/or solutions are welcome.)

10. Good tackiness to hold parts in place particularly during placement machine table movements.

11. Excellent wetting

12. Robust reflow profile

    12.1. Air or nitrogen

    12.2. Ramp to peak or ramp soak and spike

13. Minimal solder balling and beading

14. Minimal residue after reflow, cosmetically pleasing (clear is preferred.)

15. Volatile organic compounds (VOCs) compliant (mandatory), VOC-free (preferred)

16. A 6-month refrigerated life, 5-day room-temperature life

17. Minimal voiding on area array devices (<10 percent preferred and <20 percent required)

18. Probe testable by bed of nails or flying probe (NC = no cleaning, organic acid = after cleaning)

19. NC residue can be removed, if desired, by cleaning (recommend chemistry and equipment)

20. Organic acid flux can be cleaned in a deionized water (DI) aqueous wash at 140°F (maximum)

21. Paste cost comparable to industry standards

22. Minimal flux buildup in reflow ovens including exhaust ductwork

23. Material safety data sheets (MSDS) health and reactivity rating

    23.1. Rating of 2 or less is mandatory.

    23.2. Rating of 1 or less is preferred.

24. Organic acid cleaning is compatible with closed-loop DI system.

iii. Wave solder flux criteria:

1. Minimal odor

2. VOC-compliant (mandatory) or VOC-free (preferred)

3. Spray or foam application (most applications are expected to be spray)

4. Bellcore-compliant for no clean or passes Bellcore testing after aqueous wash in DI water for organic acid fluxes.

5. Robust process window

   5.1. Topside preheat = 80 to 125°C

   5.2. Wave contact time = 2 to 10 s

   5.3. Wave temperature = 261°C (same alloy as paste)

   5.4. Heavy selective soldering

6. Minimal solder balling and beading on joints and on board solder mask

7. Minimal visible residue, not sticky

8. Probe testable on a bed of nails or flying probe with NC or cleaning after organic flux

9. NC residue can be washed off if required (recommend chemistry and equipment)

10. MSDS "Health and Reactivity Rating": 2 or less is mandatory, 1 or less is preferred.

11. Organic acid flux residue is compatible with closed-loop DI water system.

6.1.7   Test Vehicle 1 (SMT Test Vehicle)

Test vehicle 1 (TV1) is shown in Fig. 6.1 and was selected for all solder paste qualifications because of the breadth of features combined with its low cost. The board is available from Heraeus Solder of Conshocken, Pennsylvania, at an approximate cost of $12 to $15 per bare board or approximately $25 to $30 for boards fully populated with dummy components from Practical Components of Anaheim, California. It is 5 in by 7.5 in, 0.062 in thick, and a single layer. A thick,

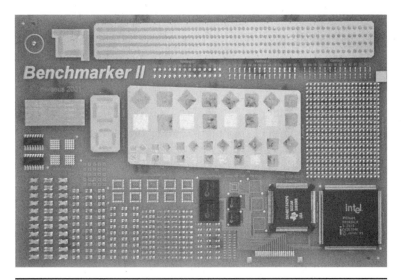

FIGURE **6.1** Test vehicle 1 (TV1) fully populated with components.

heavy board is not required for lead-free solder paste qualification, provided the reflow oven is capable of achieving the same profile for varying-thickness boards. A large number of boards were purchased and evenly divided among OSP, ENIG, and ImAg surface finishes. A tin/lead baseline and lead-free versions of the board were run in parallel with the goal of selecting a lead-free solder paste that provided performance characterization equal to or better than the tin/lead baseline. This would better ensure that the assembly yields would improve or remain the same if a product were converted from tin/lead to lead-free. Tin/lead and lead-free dummy components were ordered for the respective boards, and the tin/lead solder paste selected for the baseline was a widely used solder paste within Benchmark Electronics. All TV1 boards were reflowed once in air prior to soldering which simulated the second pass of double-sided reflow. Some surface finishes, such as OSP, degrade during thermal excursions, and it was important to qualify the solder paste, knowing it would perform well on slightly degraded surfaces. The boards were run in multiple Benchmark Electronics locations and rated by engineers and experienced inspectors acting independently. The scores from all individuals in each location were tabulated to obtain the final rating.

6.1.8   Test Vehicle 2 (Through-Hole Test Vehicle)

Test vehicle 2 (TV2) is shown in Fig. 6.2 and was selected for through-hole development and rework validation. It is a semicustom board that was designed and built by Topline Inc. of Garden Grove, California. It measures 5.5 in by 4.0 in, has 12 layers, and is 0.097 in thick. The outer

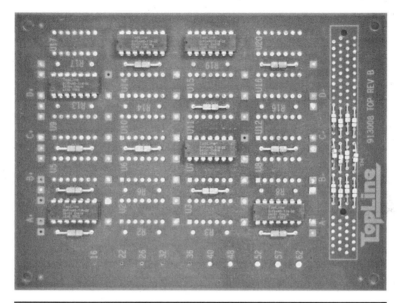

FIGURE 6.2 Test vehicle 2 (TV2) populated with through-hole components.

layers were 1 oz copper, and the inner layers were ½ oz copper, and selected components were connected to all 12 layers, which made them difficult to solder and rework. The board provided a good approximation of real product whereas a two-layer board would have been too easy to solder and unrepresentative of a real product. Components were ordered from a dummy component supplier, and the cost of the populated boards was approximately $55 per card. Similar to TV1, the boards were evenly divided between OSP, ENIG, and ImAg surface finishes. The TV2 boards were all reflowed twice in air prior to wave soldering to simulate the double-sided SMT assembly followed by through-hole assembly. All TV2 boards were soldered at a central location with engineers from multiple Benchmark Electronics locations present to rate the boards. The ratings from all engineers were tabulated for the final score.

Interconnect stress test (IST) coupons were included with TV2 and provided the opportunity to validate the integrity of the PCB after lead-free reflow. A typical coupon is the GT40800A coupon, and one is shown in Fig. 6.3, courtesy of PWB Solutions of Ottawa, Ontario. The IST coupons were separated from TV2 prior to assembly and are not shown in Fig. 6.2. The coupons were subjected to 0, 3, or 6 thermal reflow cycles with a peak temperature of 260°C at Benchmark Electronics. They were then sent to PWB Solutions of Ottawa, Ontario for electrical testing per the following criteria:

- 500 cycles (3 min per cycle)
- Power on until temperature reaches 150°C

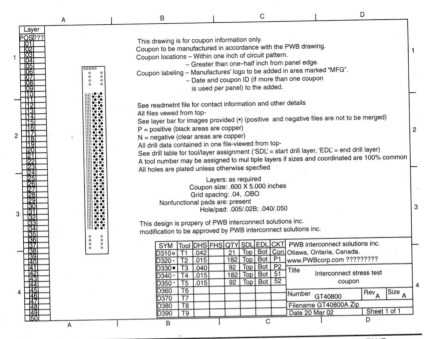

Figure 6.3 GT40800A interconnect stress test (IST) coupon. (Courtsey PWB Solutions of Ottawa, Ontario.)

- Power off until boards cool to room temperature
- Monitoring of small instead of large vias, since both cannot be tested simultaneously

Based on the results of the test, an assessment was made as to the compatibility of the PCB for lead-free assembly. A successful test provides a much higher confidence level in the long-term reliability of the PCB after the higher temperatures associated with lead-free assembly. Refer to Chap. 8 for more discussion of the IST test.

Although TV2 was a 100 percent through-hole test vehicle, all boards were subjected to a double reflow in air prior to lead-free wave soldering. This was done to simulate a board with double-sided SMT and through-hole with an OSP finish which degrade with each reflow operation.

6.1.9 Lead-Free Paste Qualification

Major solder paste vendors were asked to submit SAC305 (preferred) or SAC396 (acceptable) based on the requirements outlined above. The SAC396 is the SAC alloy with 95.5Sn3.9Ag.6Cu and, at the time of this development activity, considered a possible industry standard. The NCMS work noted in Sec. 6.1.2 indicated the SAC305 and SAC396 could be considered equivalent. Nine no-clean and eight organic acid solder pastes were made available, and a legacy NC tin/lead solder paste was selected as the tin/lead baseline. Similar testing was conducted on organic acid but not recorded here due to space constraints.

6.1.10   Laboratory Analysis of the No-Clean Paste

The initial comparisons of the lead-free pastes began with laboratory analysis, to reduce the number of candidate lead-free pastes. The viscosity material property was measured, and three tests were conducted: slump, solder balling, and wetting. The tests were based on IPC criteria for tin/lead, but with temperatures elevated 30°C from tin/lead criteria, since the lead-free criteria had not been released by IPC at that time. The summary of the laboratory results appears in Table 6.1.

The tin/lead baseline was the top performer, especially in the wetting test where a lower wetting time indicated better wetting which is desired. The results were consistent with previous work that indicated the lead-free alloy did not wet or spread as well as tin/lead. However, there was no justification to remove any lead-free paste based on these results.

A number of test vehicle 1 boards were provided to multiple sites for the evaluation, with each site performing the identical tests independently. This would provide a higher confidence level in the

No-Clean Solder Paste	Viscosity (cP)	Slump (no. of space violations)	Solder Balling (1 = good, 4 = bad)	Wettings
A	2570	0	2	48
B	1930	11	1	91
C	1611	4	2	41
D	2380	0	3	95
E	2170	3	2	39
F	1359	0	3	112
G	2370	0	2	46
H	2170	8	2	41
I	2600	2	2	48
SnPb baseline	2150	0	2	16

TABLE **6.1**   Laboratory Results for Lead-Free Solder Pastes and Tin/Lead Baseline Solder Paste

lead-free solder paste selection if a paste performed well across different sites, since these sites afforded different atmospheric conditions
and equipment.

6.1.11   Evaluation Phases

The evaluation was divided into three successive phases:

A1 = printing (no reflow)

A2 = reflow of paste (no components)

A3 = reflow of fully populated boards

The operating parameters were held constant for all pastes. For
example, the same print speed was used for all solder pastes; and one
reflow profile was used for the tin/lead baseline solder paste, and
one reflow profile was used for all lead-free solder pastes. All reflows
were performed in air, to select a candidate paste that would also be
cost-effective. Reflow in Nitrogen is considerably more expensive.
The lead-free reflow profile was as follows:

- Ramp to peak at 1 to 2°C/s

- Peak temperature range = 240 to 248°C

- Time above liquidus (TAL) = 60 to 90 s

- Air atmosphere in the reflow oven

Pastes were successively eliminated in each test, which simplified
the testing and minimized the number of test vehicles and components required. For example, A1 entailed printing only, and the boards
were washed after the prints had been assessed, which allowed the
boards to be used multiple times. The boards in A2 and A3 were
reflowed, which meant they could only be used once. Moreover, A2
did not include components which minimized component consumption to minimize cost. If a paste did not print well in A1, it was dropped
from further evaluation, and the same was done for A2 and A3.

The spacing between linear apertures on TV1 varied from 300 to
120 μm. A representative print deposit for a specific spacing appears in
Fig. 6.4. The linear deposits are vertical, horizontal, and at 45°, which
allowed assessment of print quality in the respective directions. The
assessment of the print quality included objective data such as the number of bridges between the linear deposits and assessment of the print
quality in a subjective fashion. In the latter the prints were rated as a
1, 2, or 3 which corresponded to preferred, acceptable, and unacceptable. In all cases, the lower the number, the better the solder paste is
rated. Selected boards were printed after an intentional 1-h delay to
assess print performance in production environments. These boards
were given a higher relative weight during the compilation of scoring
to ensure poor performance on this test would be recognized. Selected

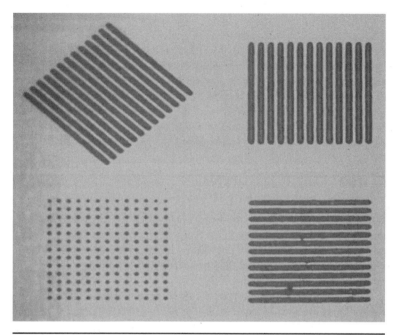

FIGURE 6.4   Representative lead-free solder paste print deposit on TV1.

boards were subjected to 30°C and 80 percent RH for several hours to simulate extreme conditions, and other boards were reflowed to a peak temperature of 185 to 190°C, which is the likely soak temperature in the reflow oven before the final step of reflow.

Figures 6.5 and 6.6 illustrate the wetting test portion of TV1 where successively less paste was printed as represented by the 27, 21, and 16 above the pad. More paste was printed on the 27 pad, and successively less was printed on the 21 and 16 pads. Figure 6.5 illustrates the best wetting that was obtained with a lead-free paste after reflow, and one can see the paste did not fully wet, or solder, the entire pad. The

FIGURE 6.5   Best lead-free wetting after reflow.

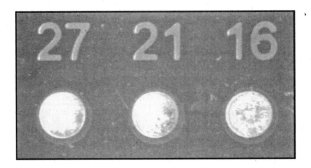

FIGURE 6.6   Tin/lead wetting after reflow.

same location on TV1 is shown in Fig. 6.6 for the tin/lead baseline paste, and one can see the significant improvements. The tin/lead paste wets almost the entire pad on the 27, 21, and 16 pads. The result was not surprising because it is known that tin/lead wets and spreads better compared with lead-free.

The complete summary of all data collected in A3 appears in Fig. 6.7. Evaluation A3 represents the reflow of fully populated TV1 boards and only includes pastes that were not removed from consideration during A1 and A2. In all cases, a lower number indicates a better-performing solder paste.

The data description in Fig. 6.7 is as follows:

1. ICT probe ability: number of no contacts during flying probe electrical continuity checks.

2. Solder balls: number of solder balls found on 1206, 0805, 0603, and 0402 resistors and capacitors.

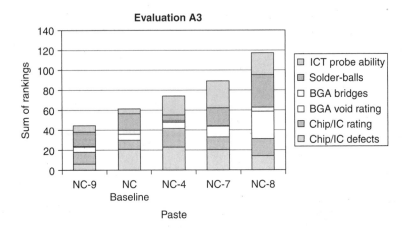

FIGURE 6.7   Summary of the data for the no-clean solder paste in evaluation A3 on test vehicle 1.

3. BGA bridging: number of bridges on all components including BGAs.

4. BGA voiding: voiding in BGAs as measured by transmissive X-ray.

   a. 1 = 10 percent or less by area

   b. 2 = 11 to 19 percent by area

   c. 3 = >20 percent by area

5. Chip IC rating: subjective rating (1, 2, or 3) of the wetting on components with leads of terminations (SOIC 14, 0805, etc.). The solder joint volume and wetting angles are key attributes of this inspection.

6. Chip IC defects: number of defects found on 1206, 0805, 0603, 0402 resistors and capacitors. Defects include tombstones, lifted components, and misregistered components.

A weighted scale was applied to the data. For example, a solder open was given a higher weight as compared with a solder bridge because the former is more difficult to detect electrically and visually. NC-9 is the top-performing lead-free no-clean (NC) paste, which outperformed the tin/lead baseline (NC baseline), which was not expected given the laboratory results indicating the lead-free pastes did not wet as well as the tin/lead baseline. The results are positive because it indicates that if a product were converted from tin/lead to lead-free, the SMT assembly yields would likely remain the same or improve. NC-9 is the lead-free NC paste selected for assembly at Benchmark Electronics.

6.1.12 Mechanical Validation of the Solder Joint
Additional validation tests for the lead-free solder pastes were conduced including cross sections, BGA ball shear testing, and component lead pull testing. The results for lead pull testing appear in Fig. 6.8 and include the minimum, maximum, and mean values. NC-9 is the top-performing lead-free NC paste, and it performed well relative to tin/lead NC baseline, and the additional tests mentioned earlier. The mechanical validation test results provide a higher confidence level in the integrity of NC-9 relative to the tin/lead baseline.

6.1.13 Reliability Comparison between Tin/Lead and Lead-Free
A final test was run on TV1 comparing lead-free boards assembled with the top-performing lead-free solder pastes with the tin/lead baseline in *highly accelerated life testing* (HALT). This is a 24-h test combining vibration and temperature cycling with continuous electrical monitoring of components and solder joints. Although it is not possible to make long-term reliability predictions with HALT, it does

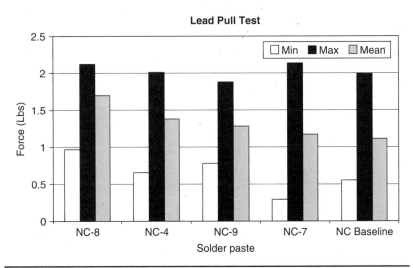

FIGURE 6.8   Lead pull test results for no-clean solder paste.

allow for a relative reliability comparison between the tin/lead baseline and lead-free test vehicles. This test is cost-effective and complements extensive thermal cycling testing which has been done previously in the industry. The HALT parameters are shown in Table 6.2. The test is 24 h long, and the temperature is cycled between 0 and 100°C, while the vibration is steadily increased.

The failure rating given in Eq. (6.1) was derived by weighting a factor to rate TV1 boards with fewer failures as better, and TV1 boards with failures that occur later in time as better. A low failure rating is best, while a higher failure rating indicates poorer performance. The formula used to calculate the failure rating is

$$\text{Failure rate} = \sum_1^n \left(\frac{n}{t_n} \right) \tag{6.1}$$

where n = failure number and t_n = time of failure.

Description	Level	Time
Thermal cycle	0°C–100°C 50°C/min	13-min Dwell (h)
Vibration Thermal cycles 1–16	10 g	8
Thermal cycles 17–32	15 g	8
Thermal cycles 33–40	20 g	4
Thermal cycles 41–48	25 g	4

TABLE 6.2   HALT Test Parameters for TV1

Of specific interest in this study is the top-performing NC paste and board finish. The best-performing paste was NC-9, compared to the worst-performing paste NC-4 and the tin/lead baseline of 0.57. A similar set of results was derived based on board surface finish during the HALT, with the lower the number, the better. The ENIG (0.50) outperformed the ImAg (0.64) and OSP (0.87) surface finishes.

6.1.14   Lead-Free Wave Solder Flux Qualification

Major flux vendors were asked to submit fluxes based on the criteria outlined in the lead-free requirement section above. Evaluations B1 and B2 were run to select the top-performing liquid fluxes. Some fluxes were did not perform well in B1 and were removed from consideration. All TV2 boards were wave soldered on a lead-free (SAC305) wave solder with identical parameters for wave solder temperature, board preheat, conveyor speed, and wave height in B1. Some modifications of the parameters were made for B2, and all boards in B2 were soldered with the same parameters.

The solder temperature range was 280 to 285°C, and conveyor speeds were between 1.5 and 3 ft/min. Topside board preheat was set to a range of 120 to 130°C. The parameters tested for the wave solder qualifications were as follows:

1. *Odor.* This represents the smell of the flux on the board after soldering with 1, 2, and 3 corresponding to preferred, acceptable, and unacceptable.

2. *Residue.* Rating of the residue left on the board with 1, 2, and 3 corresponding to preferred, acceptable, and unacceptable. A light clear residue would be given a 1, a slightly heavier and darker residue would be given a 2, and very dark, heavy residue would be given a 3.

3. *Solder ball.* Number of solder balls on the bottom and top of the board.

4. *Solder fill.* Number of through holes that did not have 90 percent or greater flow through to side 1. (*Note:* IPC class 3 criteria require 75 percent, but 90 percent was chosen for ease and speed of inspection.)

In all cases, a lower number indicates a better-performing wave solder flux. Liquid flux NC-1 is the top-performing lead-free wave solder NC flux, for all surface finishes, with the majority of the defects based on incomplete solder fills. When the same data were arranged by surface finish, OSP was the worst-performing PCB finish, across all solder fluxes. NC-1 performed best on OSP as compared with the other fluxes. It is possible to argue that fluxes NC-2, NC-3, and NC-4 are equivalent if ImAg or ENIG were the only surface finishes selected. However, once OSP boards are taken into account, NC-1 is the top performer.

6.1.15 Benchmark Electronics TV3

A third test vehicle, referred to as the Benchmark Electronics test vehicle 3 (TV3), was designed and assembled. The objective was to demonstrate that the lead-free formulations, alloys, and experience identified with TV1 and TV2 were sufficiently understood and developed. Benchmark Electronics TV3 definitively showed that the knowledge and expertise gained with TV1 and TV2 would allow for a successful transition into manufacturing of a highly complex board assembly using lead-free material and processing techniques throughout the company. The topside view of TV3 appears in Fig. 6.9.

The design of the Benchmark Electronics test vehicle 3 integrated various component designs and complexities that best simulated current and future customer design concepts. The board design incorporated elements to accommodate large board thickness, four-point bend testing, highly accelerated life testing (HALT), highly accelerated stress screening (HASS), and in-circuit test (ICT) on various board surface finishes, as well as special requests from both internal and external customers. This was done to find equipment and processing limitations so that corrective actions or process improvements could be instituted before the mandatory deadlines for lead-free implementation.

Summary of the Benchmark Electronics TV3

- 11 in by 16 in.
- 0.150 in thick.
- Double-sided SMT with through-hole components.

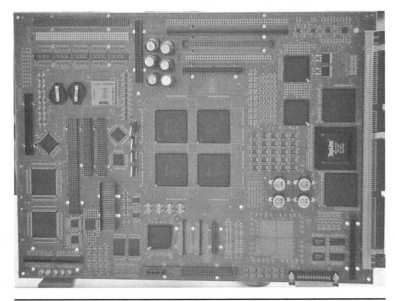

FIGURE 6.9   Topside view of the Benchmark Electronics test vehicle 3.

- 16 layers alternated between signal and ground plane layers.

- High glass transition temperature (Tg > 175°C).

- Blind and buried vias incorporated into the artwork. Thermal relief patterns incorporated into the artwork and through-hole areas. The thermal relief pattern is applicable for internal ground planes and, to a lesser extent, signal planes. Ground planes are normally thicker than signal planes and act as heat sinks which can make through-hole soldering more difficult. Specifically, the heat sink can cause the molten solder to solidify prior to reaching the topside of the board, which is not desired. A common thermal relief pattern is the wagon wheel which consists of four spokes emanating from the through hole to the applicable ground plane as contrasted with the ground plane making full contact with the through hole. The four spokes dramatically reduce the rate of heat transfer from the through hole to the ground plane.

- OSP, ImAg, and ENIG surface finishes.

- Tin/lead (HASL) surface finish for the tin/lead baseline boards.

- IST coupon.

- Various other features.

Approximately 80 boards were assembled, including a small number of tin/lead baseline boards. The boards were subjected to 100 percent visual inspection, automated optical inspection, X-ray inspection, ICT, forced BGA and through-hole rework, HALT, HASS, and four-point bending tests. The latter assesses BGA joint strength in response to mechanical stress.

The primary objective of validating the lead-free chemistries and alloys was met, but two areas requiring further development were identified:

- Wave solder. Optimize the process to consistently obtain class 3 (>75 percent fill) through-hole solder joints on 0.150-in-thick boards.

- Through-hole rework and minimizing copper dissolution on OSP and ImAg boards during rework. Excessive copper dissolution completely removes copper pads, which destroys the integrity of the board and results in the board being scrapped. ENIG boards have a nickel barrier over the copper layer, and the SAC305 solder is not able to penetrate the barrier and dissolve the copper. This indicated that ENIG boards are not at risk for copper dissolution during rework.

The greatest benefit of TV3 was the assurance it provided to existing and prospective customers that the lead-free process developed with TV1 and TV2 was capable of producing complex boards.

6.1.16 TURI TV3

Benchmark Electronics joined the New England Lead-Free Electronics Consortium—see http://www.turi.org/industry/research/supply_chain_program/electronics which is managed by the Toxic Use Reduction Institute (TURI) based at the University of Massachusetts in Lowell. The consortium included academia, industry, and government and offered a number of key benefits:

1. It was low-cost. Most members primarily made in-kind contributions with additional funding by the EPA.

2. The members were primarily located in New England, and consensus was achieved in a timely manner.

3. It enabled better understanding of the previous lead-free work done by the consortium and access to their expertise.

4. It offered involvement with all parties affected by lead-free compliance including PCB and component suppliers, OEMs, academia, and the EPA, which provided a better understanding of their respective areas of concern. A successful lead-free solution requires good teamwork and communication throughout the supply chain.

Benchmark Electronics assembled the consortium's test vehicle 3, which is referred to as the TURI TV3. The boards included a number of active components donated by various consortium members which was valuable because TV1, TV2, and the Benchmark Electronics TV3 were assembled primarily with dummy components. Although the dummy components provide an excellent means of ensuring the integrity of SMT and through-hole solder joints, they do not ensure the overall reliability of the board, which is contingent upon components designed for the higher temperatures associated with lead-free. A reliable product requires good solder joints, components, and PCBs, and if any one of the three aspects is compromised, the product's reliability will be compromised.

6.1.17 TURI TV3 Overview

The TURI TV3 initiative focused on examining the manufacturing issues of implementing lead-free electronics assembly for PCBs more closely resembling actual production boards. The TURI TV3 board was designed to better simulate actual production boards commonly used in industry and had a variety of component types and finishes, as well as some active circuitry that could be tested for electrical performance. The object of the TURI TV3 board was to compare the solder joint integrity and circuit performance between lead and lead-free electronics assembly for production volumes. The primary metrics for this comparison were defects per unit (DPU) as identified during the visual inspection process and peak pull strength after thermal cycling as measured during pull testing. In addition, HALT was conducted by consortium members. A total of 36 NC TURI TV3 boards were assembled in Benchmark Electronics. Of these, 24 boards were lead-free and 12 were tin/lead baseline. The NC chemistry was identified through work completed in TV1 and TV2 phases.

The TURI TV3 was comprised of 20 layers and had a large footprint of 16 in by 18 in. The PCB was densely populated with components on both sides. Each board contained 1713 SMT-type components and 53 THT-type components, for a total of approximately 62,000 components included in the experiment.

All tin/lead and lead-free through-hole soldering was done on tin/lead and lead-free solder fountain, respectively. Production lead-free wave solder and selective solder platforms were not available at the time of the build.

6.1.18   Factors and Levels of the Lead-Free Experiment

The following factors were selected for the TURI TV3 board and were based on results of the TURI phase I and II efforts, as well as the collective experience of the consortium members: 2 lead-free solder suppliers (A and B), 3 surface finishes, and 2 types of laminate materials (Y and Z), for a total of 12 experiments based on 3 factors and $3 \times 2 \times 2$ levels. The three surface finishes were ENIG, ImAg, and OSP. The lead-free solder from both suppliers used the SAC305 alloy and NC flux. Other factors such as solder reflow profile (as recommended by the supplier) and atmosphere (air) were fixed based on member consensus. Each board underwent at least two reflows during assembly, one for topside of the PWB and another for bottom side of the board.

The lead baseline consisted of 6 experiments of 2 factors (solder suppliers and surface finish) for 2×3 levels to reduce the number of iterations. Only one laminate material was used for the lead baseline experiments.

6.1.19   TURI TV3 Components

Surface mount and through-hole technology components were used for the TURI TV3. Component types include ball grid arrays (BGAs), small outline integrated circuit (SOIC), connectors, resistors, capacitors, relays, inductors, and quad flat packs (QFPs). A variety of component finishes incorporated including the following: nickel/palladium/gold, tin, tin/lead, gold, nickel/gold, tin/nickel, palladium/silver, tin/copper, tin/silver/copper, tin/bismuth, and matte tin. Some components were available in daisy-chain configurations for electrical testing.

6.1.20   TURI TV3 Thermal Cycling and HALT

The thermal cycling parameters were as follows:

- Compliant with IPC-9701, test condition 1
- 2000 cycles between 0 and 100°C
- 40 min/cycle
- Ramp rate = 10°C/min

- 10-min dwell at 100°C
- 10-min dwell at 0°C

The highly accelerated life testing parameters as follows:

- Temperature cycling = –60 to +160°C
- Vibration = static to 60 g
- Dynamic measurement of resistance: 17 daisy chains

6.1.21   TURI TV3 Results and Conclusions

The results and conclusions of the TURI TV3 phase were as follows:

1. Based on the visual inspection results, the solder joint integrity for SMT components on the lead-free boards was comparable to that of the tin/lead boards for the selected solder paste suppliers, laminate suppliers, and board surface finishes. Therefore, attaining acceptable solder joint integrity with lead-free assembly was possible using existing equipment and with careful selection of materials and close control of processes.

2. The number of lead-free through-hole defects was higher than the tin/lead through-hole solder defects. Additional optimization will be required for lead-free through-hole soldering.

3. Pull testing after thermal cycling showed no statistical differences for tin/lead or lead-free solders, component surface finishes, laminates, or PCB surface finishes.

4. HALT resulted in two failures; one was of solder joints, and the other failure mode was lead fractures which indicated the leads fractured before the joints failed. It was concluded that there were no significant differences between tin/lead and lead-free boards for HALT.

Overall, the TURI TV3 initiative provided a higher confidence level in the integrity of lead-free assembly in production volume applications.

6.1.22   TURI Test Vehicle 4 (TV4)

At the time of this writing, preparations are being made to assemble TURI's test vehicle 4 (TV4) which will focus on the following:

1. Long-term reliability of the lead-free boards in environments that may include thermal excursions in conjunction with vibration and/or shock. This environment will be used to study the boards for high reliability as in military and space applications. A good resource for additional information is the NASA/Department of Defense Lead-free Electronics Consortium which can be found at

http://www.teerm.nasa.gov/projects/NASA_DODLeadFree Electronics_Proj2.html

2. Forced rework of BGAs and through-hole components with the latter focusing on minimizing copper dissolution and thermal degradation of the PCBs. New rework techniques will be evaluated including a unique solder fountain nozzle designed to minimize copper dissolution.

3. Halogen-free PCBs. Halogens are typically fire retardants integrated into the PCB. They are not currently banned, but there is a potential that their use will be restricted in the future by legislation or market forces. Refer to Chap. 8 for more details on halogen-free laminates.

4. Alternative surface finishes including lead-free HASL utilizing the Sn100C lead-free alloy and possibly a new surface finish incorporating nanotechnology.

5. Use of the Sn100C alloy offered by Nihon Superior for through-hole soldering and rework. Its performance will be compared with the SAC305 on identical through-hole soldering and rework platforms. The consortium members for the TURI TV4 initiative appear in Fig. 6.10.

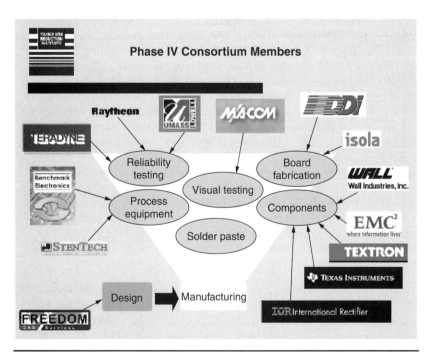

FIGURE 6.10   New England Lead-free Electronics Assembly Consortium members for the design, assembly, and test of the TURI test vehicle 4 (TURI TV4).

6.2 Communications with the Design Teams and Material Supply Chain

Design for the environment (DfE) guidelines have become a dynamic communication process with the design teams. DfE guidelines have been developed to date and continue to be updated given new information and analysis learned. Two main drivers have evolved for lead-free design guidelines: the initial driver is the RoHS directive requirements and related banned substances, and the second is the increased temperatures required for lead-free manufacturing. The following DfE guidelines relate to lead-free assemblies and the manufacturing of green electronics.

6.2.1 Design of PCB Material

The PCB components must be free of all six banned RoHS substances, with lead being the primary banned substance which has historically been used for years in electronics manufacturing. Lead-free material, including SMT and through-hole components and PCBs, must be designed to withstand the additional temperature requirements of lead-free manufacturing. The maximum case temperature requirement for SMT components must be rated to 260°C. Special attention should be made when selecting components and verification of the requirement in the component data sheet. Some SMT components, including BGAs, have been discovered to have reflow temperatures under 260°C (250°C was the specific case). This circumstance will cause manufacturing capability problems as well as possible reliability issues, given the lead-free peak reflow temperature range of 240 to 248°C and process tolerances of the related equipment. If the lead-free component selected has availability issues and cannot be procured until a later date, this will create manufacturing process concerns, assuming that the assembly must be built and the leaded or non-RoHS-compliant version is delivered. This situation is described as a hybrid assembly and cannot be sold as fully RoHS-compliant. Hybrid assemblies are discussed in detail later in the chapter. Given the various standard pad geometry and hole size requirements, the DfE impact for the assembly process is minimal.

When one is qualifying PCB material, many factors should be analyzed and reviewed. Refer to Chap. 8 for more information on PCB materials. The following are some industry guidelines for lead-free manufacturing and some practical lessons learned to date:

1. The finished product must be compliant with the RoHS directive and free of all six banned substances. Three of the six banned substances have been historically used in the industry—lead and the two banned flame retardants, PBB and PBDE.

2. The PCB must withstand two temperature guidelines, the first being able to be processed at the 260°C lead-free thermal

reflow temperature and the second requirement being a minimum glass transition temperature T_g of 170°C. The T_g is defined as the maximum allowable temperature a board can see without degrading the FR4 material or raw PCB material. This T_g has been increased from the historical 140 to 170°C to withstand the higher temperatures of lead-free soldering. During selection of the laminate material, review and analyze all the applicable reliability data in the industry and from your PCB supplier, as many laminates are being improved given the learning curve of lead-free soldering. The selection of laminate material should be applicable to the complexity of the design, the end application of the product, and the temperatures required during manufacturing. If a product is complex and requires blind vias, a particular laminate material may not be favorable.

3. The three most used surface finishes for lead-free are OSP, ENIG, and ImAg. The selection is dependent on factors such as quality, reliability, and cost. One emerging surface finish solder is lead-free HASL. This finish has been improved by correcting a recurring problem of pad level coplanarity, as shown by some test vehicles to date. If a higher reliability level is desired, research and analysis of newer finish technologies can be conducted.

6.2.2 Component Supplier Readiness Methods

Given the industry transition and the requirements for lead-free manufacturing, supplier readiness surveys and communication of supply chain requirements should be conducted. Readiness will be varied given the infrastructure of the supplier and end product design and manufacturing requirements. The initial communication to the supply base is critical to understand the suppliers' lead-free and RoHS transition road map. The following steps are guidelines to survey, communicate, and collect the information needed to transition and design a lead-free product.

1. Is the supplier knowledgeable of the compliance requirements, including the maximum concentration limits and requirements (i.e., homogenous material, see Sec. 6.4.4) and compliance effective date?

2. How will the product be labeled or marked to indicate compliance, and is the manufacturer's part number (MPN) being changed to reflect compliance or is a date-code transition being introduced? The packaging or component may have the applicable compliance labeling; this may be dependent on the product, size, and specific supplier process.

3. Will the supplier provide a certificate of compliance (C of C) or material declaration for the product? The declaration is required under the RoHS due diligence requirement for a compliance assurance system. The compliance documentation will be in varied formats, as a set standard has not been fully deployed in the industry. Request of specific processing information should be made such as maximum case temperature, moisture sensitivity information, and copies of reliability testing completed. With lead-free manufacturing, suppliers must have their components rated to 260°C given the SMT reflow processing temperature of lead-free solder. This change in temperature has also changed moisture level requirements of components (see Chap. 7 for more details).

4. What are the transition strategy and obsolescence process? Is the specific component in question being changed to become compliant? Or is the supplier not planning to be compliant given the cost and infrastructure?

The example in Table 6.3 is a generic certificate of compliance which could be used for electronic components, assemblies, or systems.

6.2.3   Component Supplier Survey Results

A data collection and analysis process is recommended to obtain the readiness information, given the process development impacts and build implications if a component is not compliant now or will

---

RoHS Certificate of Compliance

Company Inc. hereby certifies to the best of our knowledge that the product specified below conform to the Restriction of Hazardous Substances (RoHS) directive EU 2002/95/EC.

Company Inc. declares that the RoHS-compliant product below is exempt per the applicable exemption listed and permitted by RoHS directive 2002/95/EC.

Part number: _____

Description: _____

Exemption declared: _____

_____   _____

   Authorized Signature   Date

   _____

   Company Inc. Production Site

TABLE 6.3   Generic Certificate of Compliance (C of C)

Part Number	RoHS-Compliant Currently?	RoHS-Compliant Effectivity?	New RoHS Part Number to Be Issued?	RoHS Date Code Transition?
	Yes / No	Date	Yes / No	Date

TABLE 6.4   Generic Component Supplier Survey

become compliant in the future. This will require redesigns and/or investigation for additional supplier sources. Surveys which were distributed to the supply base in late 2005 resulted in various compliant road maps depending on the component types. Many of the common passive (resistors, capacitors) and active (integrated circuits) components were compliant or provided a plan for compliance with noncompliant components being available until further notice. The specialty suppliers, such as of mechanicals, cables, and power components, responded with varied long-term compliance plans. The product labeling and compliant documentation were also varied depending on the manufacturer, with most suppliers changing part numbers for ease of control and the remaining suppliers incorporating a date code transition to compliance. Material declarations were available upon request with a small percentage of manufacturers having the documentation via their websites. This percentage is increasing with time. Table 6.4 is a typical component supplier survey form and will complement the questions in Sec. 6.2.2.

6.3   Recommended Steps for Successful Lead-Free Conversion (Process and Product)

The steps below are recommended for successful deployment of lead-free assembly at various manufacturing locations following the work outlined in Secs. 6.1 and 6.2. Discussions include the recommended equipment for lead-free assembly, recommended steps to qualify a process and product for lead-free assembly, and design for manufacturing (DfM) changes.

6.3.1   Manufacturing Site Process Qualification

Site process qualification includes assembling test vehicles such as TV1 and TV2 which are the SMT and PTH test vehicles, respectively.

A key attribute is their low cost which makes the qualification cost-effective. The qualification process should proceed as follows:

1. Schedule a corporate lead-free presentation for the management team, manufacturing operations, and applicable personnel from each functional area to raise awareness and communicate status of the initiative. Two hours should be allocated, and the presentation can be made on-site or via a conference call. The training emphasizes that lead-free affects the entire process including design, component acquisition, kitting, assembly, test, rework, shipping, field service returns, and adequate recordkeeping. The last is required to respond to requests from customers and/or regulatory authorities regarding the lead-free (RoHS compliance) status of products. It is insufficient to simply place an "RoHS-compliant" label on a completed assembly. The recordkeeping provides the underlying justification and can include C of C for components used on the product and sample X-ray fluorescence (XRF) to verify RoHS compliance which is discussed in detail later in this chapter.

2. Surface mount assembly lines will likely have to be switched between tin/lead and lead-free as required, whereas the wave solder, selective solder, and rework fountains should be dedicated to lead-free because it is not practical to interchange the tin/lead and lead-free molten solder. Some equipment manufacturers allow for tin/lead and lead-free pots to be rolled into the wave solder as required, and companies may elect to consider this option.

3. The training emphasizes the importance of process controls to minimize the chance of cross-contamination between tin/lead and lead-free assembly processes which could include using a lead-free paste on a tin/lead product or placing tin/lead bar solder into a lead-free wave solder.

4. Appoint a project manager and site champion as the focal point for the RoHS deployment.

5. Survey the customer base to determine respective road maps for conversion. Maintain a matrix of survey responses and immediate interest in lead-free services by customers. Customer presentations are also recommended and can be scheduled as required. Anticipate customer lead-free requirements in order to respond promptly to their requests.

6. Reference the lead-free equipment recommendation below for capital planning. Complete a lead-free equipment gap analysis, and contact the applicable equipment manufacturer to proceed with the capital request process for noncompliant equipment. All equipment ordered in the future should be lead-free-compatible.

Lead-Free Equipment Recommendations

Assembly and test equipment that is not involved in thermal excursions can be considered lead-free-compatible. This would include incoming inspection equipment, screen printers, stencil cleaners, placement equipment, conveyors, X-ray equipment, test platforms, X-ray inspection, automated optical inspection (AOI) platforms, and microscopes. The AOI platform may require programming modifications because the lead-free joints will be slightly duller and grainier than the tin/lead joints. The risk of cross-contamination between tin/lead and lead-free products on these platforms is minimal, and the products can be interchanged provided the platforms are cleaned according to a standard preventive maintenance schedule. The risk of cross-contamination between tin/lead and lead-free in the stencil cleaner is negligible, and the same stencil cleaner can be used for both alloys.

Equipment involved in thermal excursions must be lead-free-compatible according to the following guidelines. In addition, it must be rated for the maximum board size and thickness that will be processed.

1. *Reflow oven.* Forced convection ovens with 10 or more zones are recommended, since they are capable of bringing the board to a peak temperature range of 240 to 248°C in continuous operation. Six- and eight-zone ovens may be acceptable for smaller products and/or if the conveyor speed is lowered. Two to three cooling zones are recommended.

2. *Temperature recording devices.* These electronic devices pass through the oven with the assembly and are used to monitor the temperature of the board as a function of time while traveling in the oven. Consult the manufacturer to verify the insulated box containing the device is adequate to protect the circuitry.

3. *Wave solder.* The wave solder must be specially coated to ensure the SAC305 does not prematurely wear the surfaces that come in contact with the molten solder. Consult the manufacturer to confirm its compatibility. Three zones of bottom heaters and two bottom zones of top preheaters are recommended along with the maximum wave contact area available. The *contact area* is defined as total length of the solder wave that makes contact with the bottom side of the board during soldering operations. Increased contact areas improve the heat transfer and facilitate the flow of solder to the topside of the board, which is especially important for thicker boards with multiple layers. They also allow faster wave solder conveyor speeds.

4. *Selective solder.* This process should be considered for high-heat-sink through-hole applications. This platform allows the board to be placed over a fixed nozzle for a programmable

time and greater contact time as compared with the wave solder. It is different from the wave solder, where the maximum contact time is limited by the slowest conveyor speed.

5. *Solder fountain.* This is required for through-hole rework. Forced convection or infrared preheat that is capable of preheating the topside laminate of the board to 150°C is recommended. The solder pot should be coated in a similar fashion to the wave solder and be capable of temperatures up to 285°C. A large solder pot is recommended to provide improved thermal capacity.

6. *BGA rework platform.* This is required for BGA rework. It recommended to use forced convection or infrared preheater capable of preheating the topside board to 160°C. In addition, forced convection reflow of the BGA should be capable of bringing the BGA solder balls to a minimum temperature of 240°C without exceeding 260°C on the BGA package. Some manufacturers offer the vacuum solder capability which can be used to vacuum or remove residual solder from the BGA pads after the BGA has been removed. The vacuum option can also be used to vacuum solder from through-hole components, and this option is recommended.

7. *Solder irons.* These must be lead-free-compatible and include larger solder tips to provide more heat for selected rework applications. Consult the manufacturer as required.

8. *X-ray fluorescence handheld analyzer.* This equipment is not involved in thermal excursions but is recommended for all manufacturing sites. XRF is used for sample checks of incoming components, outgoing assemblies, and solder pot checks to verify RoHS compliance (lead-free). The equipment is discussed in detail later in this chapter.

Lead-Free Conversion Process Recommendations

The recommended steps to implement a lead-free process conversion are as follows:

1. Order or upgrade SMT equipment which is not lead-free-compatible. Note that most SMT lines are lead-free-compatible for the TV1 qualification build.

2. Order the SMT TV1 or similar material and stencils along with the two top-performing no-clean and organic acid lead-free pastes. Run the lead-free pastes against a baseline of the respective tin/lead pastes used in the facility. The objective of this work is to verify the lead-free pastes have performance metrics that are close to, or exceed, the tin/lead metrics along with training applicable individuals in lead-free SMT assembly

including process control to ensure the integrity of lead-free and tin/lead assembly, as well as inspection of lead-free PCBs with the latest industry standards.

The profile for the lead-free should be the one used in Sec. 6.1.11, and the tin/lead profile for TV1 should be the current one used in the facility. All TV1 boards should be reflowed once prior to assembly to simulate the second pass of double-sided reflow.

1. Send some of the tin/lead and lead-free PCBs to selected laboratories for analysis, which should include cross sections of the joints, pull testing of lead frame joints, and shear testing of BGA joints. Additional testing such as ion chromatography can be added, which is a nondestructive test to verify the overall cleanliness of the assembly. A clean board indicates a low probability of electrochemical failure in the field. A final report along with sample boards can be provided to customers upon request. In the Benchmark Electronics Company, the lead-free team evaluated the final report findings and provided approval for the site qualification. The analysis at Benchmark Electronics did not include thermal cycling which is more expensive and time-consuming. Extensive thermal cycling and other reliability tests have been conducted on the replacement SAC305 alloy within the industry and are readily accessible. Sample portions of a TV1 analysis appear in Sec. 6.3.3.

2. Order or upgrade the platforms required for lead-free PTH soldering and rework including wave and selective soldering as well as BGA and PTH rework stations.

3. Order a TV2 or similar PTH PCBs, and solder them using the top-performing organic acid and NC lead-free solder. Run parallel boards on the tin/lead soldering platform, using their current tin/lead baseline materials. The objective is to verify that the replacement lead-free solder levels are close to or exceed the tin/lead performance. In addition, training of employees in lead-free assembly and process control is important. Qualifications should include forced PTH rework on the solder fountain and forced BGA rework. The TV2 wave solder parameters should be similar to the ones used in Sec. 6.1.14. All TV2 boards should be reflowed twice, using the TV1 profile parameters prior to the boards being wave soldered to simulated double-sided SMT followed by through-hole soldering.

4. Selected baseline tin/lead and lead-free TV2 assemblies should be sent to a qualified laboratory for analysis, which will include cross sections of selected joints to verify the integrity of the joint per IPC class 3 guidelines and the absence

of thermal damage to the board, including minimal copper dissolution of copper pads. A final report along with sample boards can be provided to customers upon request. In Benchmark Electronics, the lead-free team evaluated the final report findings and provided approval for the site qualification. Reliability analysis through thermal cycling and other means could be deferred by using industry wide published research, as was done in TV1. Sample portions of a TV2 analysis appear in Sec. 6.3.4.

6.3.2 Qualification of a Product for Lead-Free Assembly

Benchmark Electronics has standardized on the emerging industry standard of SAC305 (96.5Sn3.0Ag.5Cu) for solder paste, bar solder, and wire solder. Significant work has taken place within the industry to qualify this alloy as a replacement for the 63Sn37Pb solder alloy. However, it is important to qualify a lead-free product to ensure the solder joints and components (SMT, through-hole, and PCB) meet or exceed all requirements for the application of the product.

The development work with all test vehicles (TV1, TV2, and TV3) indicated that significant DFM changes are not necessary for SMT when converting a product from tin/lead to lead-free. Equivalent results were obtained when the boards were run as tin/lead or lead-free, using the same screen print stencil.

Lead-free through-hole soldering is more challenging than tin/lead, and DFM changes should include the following:

1. Minimize the heat sink effect on the through-hole joints by incorporating thermal stress relief patterns in signal and especially ground layers.

2. Move side 2 (bottom side) SMT components farther away from the through-hole joints to allow the use of larger nozzles for selective soldering or large shielded pallet openings for wave soldering. This will improve the heat transfer, which in turn facilitates the flow of solder into the through holes and increases the required z axis fill (75 percent for IPC class 3 joints). A minimum clearance of 0.200 in between the outside edge of all through-hole capture pads and the nearest SMT component is recommended.

3. Request IPC class 3 copper plating thickness criteria for all through-hole surfaces even if the product is class 2 or 1. This provides a more robust soldering surface to counteract the copper dissolution (erosion) which occurs during the original assembly and rework. Class 3 requires a minimum copper thickness of 0.001 in versus 0.0008 and 0.0006 in for class 2 and 1, respectively. This recommendation is especially important for PCBs with OSP and ImAg surface finishes which are most vulnerable to copper dissolution. PCBs with ENIG surface finish have negligible risk of copper dissolution because

the nickel provides a barrier between the SAC305 solder and the underlying copper layer.

4. Increase the radial contact distance of the bottom side capture pads. Experimentation has shown a 12-mil radial contact distance is more robust than a 6.6-mil radial contact. Refer to Sec. 6.6.8 for more information.

The following steps are recommended for qualification of a lead-free (RoHS-compliant) product.

a. Select a product that is representative in terms of the supply chain, component/PCB technology, and application.

b. Conduct a supply chain analysis to verify all components are RoHS-compliant. This may highlight issues in the supply chain such as

(1) Vendor readiness

(2) Unacceptable temperature ratings (260°C is required)

(3) Unexpected presence of cadmium, mercury, and hexavalent chromium

(4) Moisture sensitivity levels (MSLs) that drop by one to two levels. A one- to two-level drop in the MSL would typically mean a component could only be exposed to the environment for 24 to 96 h versus one week which makes moisture control of the components more difficult during assembly. For more information, consult Sec. 7.5.1.

1. Nonpreferred component surface finishes (tin whisker risk or solderability issues)

2. Select a PCB laminate and surface finish. While not all-inclusive, it is recommended that the following laminates and surface finishes would be acceptable for lead-free processing:

   • Laminates: Isola (formerly Polyclad) 370 HR, Isola 410, Isola 415, and Isola 620

   • Surface finishes: OSP, ImAg, ENIG, and lead-free HASL

Note that proper moisture control procedures are strongly recommended for laminate material susceptible to moisture entrapment. The controls must be in place during the fabrication of the PCB and during assembly at the EMS provider. Improper moisture control procedures can result in delamination (see Chaps. 3 and 7) occurring during assembly on the first reflow, second reflow, through-hole wave soldering, or rework. Vapor pressure during a lead-free profile is significantly higher compared with a tin/lead profile due to the higher reflow temperatures.

For an average complexity PCB, a build quantity of 30 is recommended. The surface finish can be constant for all 30 boards or evenly divided among OSP, ImAg, and ENIG. Each surface finish (and laminate) has advantages and disadvantages, and the selection should be based on component technology and product application. The following data should be collected and compared with the tin/lead version of the assembly (if applicable):

- Inspection yields including visual inspection, AOI, and X-ray
- ICT, flying probe, and functional test results

Note that development work with test vehicles indicated the same stencil can be used for tin/lead or lead-free. However, the lead-free paste does not wet or spread as well as compared with tin/lead especially on OSP and ImAg PCB surface finishes. Slight increases in the stencil aperture length can be made if it is desired to fully cover the OSP and ImAg PCB board surface finish with lead-free solder after reflow. This in not required for class 3 joints, but some customers may request complete coverage, and it can also facilitate future rework by converting the OSP or ImAg surface finish to a SAC305 HASL finish which is more robust in terms of minimizing oxidation.

7. Forced rework of a through-hole component and BGA is recommended to validate the rework process.

8. Select from the following tests as appropriate below. It is strongly recommended the lead-free assembly be subjected to the same tests used for the tin/lead assembly, which may include additional test(s) to those recommended below. Ideally, the performance of the lead-free assembly should meet or exceed the performance of the tin/lead version.

Test	Criteria Assessed
1. Pull tests of representative SMT components	Solder joint strength
2. Shear tests of BGAs	BGA joint strength
3. Cross sections of SMT and through-hole	Solder joint shape/voids
4. Scanning electron microscopy (SEM) or EDS	Solder joint metallurgy (IMC)
5. Ion chromatography	Ionic levels, cleanliness
6. SIR testing of SMT, through-hole, and rework (chemistry/requires separate run of SIR boards)	Long-term reliability

7. Thermal cycling per IPC 9701	(CTE) Mismatch issues
8. Mechanical shock, vibration, or HALT	Product specific applications validate PCB surface finish
9. Thermoire analysis	BGA dimensional stability
10. Strain testing on BGAs per IPC 9702	Strain levels at assembly/test

The above qualification steps are guidelines only, and the company is responsible for the overall reliability of the product which includes ensuring that the solder alloy, PCB laminate, PCB surface finish, and components are compatible with the product's application.

6.3.3   TV1 Laboratory Analysis (SMT Process Qualification)

The following are representative results, shown in Fig. 6.11, from evaluation A3 for TV1 SMT process qualifications undertaken at various Benchmark Electronics manufacturing sites. Evaluation A3 entails printing solder paste, placing components, and reflowing the boards. The material is representative only and allows a comparison between the tin/lead baseline and the various lead-free pastes that were evaluated. All data were collected for NC paste evaluations, and the same work was completed for organic acid solder paste but is omitted in this chapter. NC baseline is the tin/lead baseline solder NC paste, and NC vendors A, B, C, D, E, and F are the NC lead-free pastes from major paste manufacturers. The top-performing lead-free NC paste is NC-A, although it may have a different reference number from earlier qualifications.

In all cases, the lower the number in Fig. 6.11, the better the solder paste performed. NC vendor A is the top-performing lead-free NC paste, and it outperformed NC tin/lead baseline. NC vendors B through F are the remaining lead-free NC pastes that were evaluated,

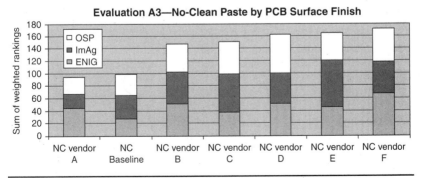

Evaluation A3—No-Clean Paste by PCB Surface Finish

FIGURE **6.11**   Visual and X-ray inspection results for TV1, evaluation A3.

and they underperformed the tin/lead baseline. Relative comparisons based on board surface finishes (OSP, ENIG, or ImAg) can also be made. For example, the NC tin/lead baseline outperformed vendor NC A on the ENIG surface finish only.

Solder joints that were cross-sectioned in this experiment showed varied results. Acceptable but increased voiding in SOIC was seen on some lead-free pastes (NC vendors A, B, and C) as compared with the NC tin/lead baseline. Voiding was most pronounced on boards with the ImAg surface finish, and this is viewed as more of a surface finish issue than a solder paste-specific issue. In addition, portions of the OSP pads are exposed primarily because the lead-free paste does not wet or spread as well as tin/lead. The exposed OSP can be soldered acceptably by slightly enlarging the stencil aperture along the length of the aperture only.

Increased voids were seen on the lead-free cross sections in 1.0-mm pitch BGA cross sections for ENIG, ImAg, and OSP board surface finishes, as shown in Fig. 6.12; and similar to the SOIC cross sections, the voiding is greatest on boards with the ImAg surface finish. All voiding shown by cross-sectioning is acceptable, but NC vendor A and B solder pastes were not selected for production. NC vendor C is the top-performing lead-free NC paste. The cross sections are not symmetrical because the pad on the BGA was smaller than the pad on TV1.

Some lead-free pastes (NC vendors A, B, and C) had lower pull test readings as compared with the NC tin/lead baseline, but all readings are high based on historical data and are considered acceptable.

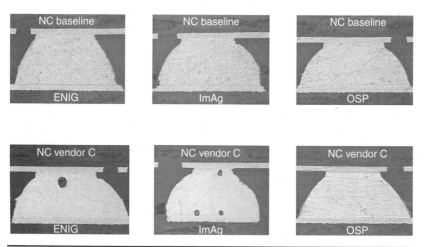

Figure 6.12   The 1.0-mm pitch BGA cross sections from TV1 for ENIG, ImAg, and OSP board surface finishes.

6.3.4 TV2 Laboratory Analysis (Through-Hole Process Qualification)

The following are representative results from TV2 PTH process qual-ifications at various manufacturing sites. All lead-free soldering was done on a selective solder or wave solder platforms using the SAC305 alloy at an approximate temperature of 280°C. Tin/lead soldering was done on the wave solder or selective solder platform with the tin/lead alloy at an approximate temperature of 245 to 260°C depending on the tin/lead wave solder configuration at the respective manufac-turing site. All boards were fluxed and preheated prior to soldering. Cross sections were made of the dual-in-position component with 14 leads (DIP14) and selected two leaded axial components. Selected DIP14 locations were connected to all 12 layers on TV2 which made them especially difficult to solder. This would normally be consid-ered a DFM violation, but it ensured that defects would be induced that allowed relative comparisons to be made.

The results of visual inspection of all wave soldered joints are shown in Fig. 6.13. In all cases, the lower the number, the better the performance of the flux or the board surface finish. Three different fluxes were evaluated, and flux A is the top-performing flux, based on the total performance for all three surface finishes. It is also evi-dent that the number of defects is highest on OSP boards as compared with ENIG or ImAg boards. Cross-sectioning of the joints showed excellent flow-through of solder to the topside of the boards and meets IPC 610 class 3 criteria which require a minimum flow-through of 75 percent. There is no evidence of thermal degradation to the board, and both A and B fluxes performed equivalently. Voiding was also minimal. These boards were soldered with the chip wave acti-vated (on). The chip wave, which is a smaller wave directly in front of the main wave, provides additional contact time between the molten solder and the board. It may be on or off depending on the PCB being soldered.

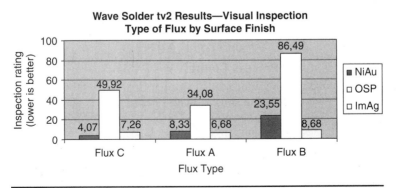

Figure 6.13 Visual inspection results for the lead-free wave soldering of TV2.

The temperature of the tin/lead molten solder was 245°C, and the joints showed excellent flow-through of solder to the topside and meet IPC 610 class 3 criteria which require a minimum flow-through of 75 percent. There is no evidence of thermal degradation, and voiding is nonexistent. Under controlled conditions, similar results were achieved with lead-free.

Some cross-sectioned joints were not acceptable. The chip wave was turned off for these samples, and this action in conjunction with the fact the boards had an OSP surface finish resulted in the significant reduction in hole fill. OSP is more difficult to solder than ENIG and ImAg.

6.4   Managing the Mix of Lead and Lead-Free Component Finishes Including Hybrid Processing Methods

Given the industry transition of material to RoHS compliance and the many exemptions at the product, component, and technology levels, the mixture of lead and lead-free components, assemblies, and systems may be prevalent in many manufacturing facilities. The requirement for process control becomes critical with so many opportunities for error. Process control is one aspect which supports the RoHS request for a compliance assurance system or due diligence controls. Section 6.2 focused on the communication of requirements and applicable data collection that the suppliers' processes and product must be compliant. The focus of this section is the compliance of manufacturing process areas, associated risk, and subsequent control to mitigate such risk. Deployment of a compliance control program provides internal detection and containment of noncompliance. When nonconformances materials are found in the manufacturing process, they can be corrected promptly. This system is costly to implement, but offers less overall cost when compared with the consequences of escape and detection by a RoHS compliance authority. The following section discusses in detail a due diligence system including material and manufacturing process control, XRF verification, and the managing of hybrid processes and assemblies.

6.4.1   Material Process Control

The first step of a manufacturing compliance system is the control of material given the varied stages of compliance from supplier to supplier and commodity to commodity. The lead-free transition has been, and continues to be, challenging with all the various exemptions, such as the "lead in solder" (also known as RoHS-5); high-lead-content spheres are exempt in some components including various ceramic ball grid arrays (CBGAs) and lead as an alloying element in metals such as hardware and mechanical components. The goal is to confirm compliance and segregate the material. Compliance can be achieved through many forms including receipt and storage of the C of C or material declarations from the supplier. The latter is a document which will declare compliance for the specific manufacturer's part number. The second process to verify compliance is XRF verification, which is discussed in detail later in this section.

The immediate goal is to create a method to identify, segregate, and store material through the entire manufacturing process including receiving, stockroom, manufacturing floor, and field returns. The deployment of such a process will prevent mixing of material at the component and assembly levels while substantiating the compliance assurance system. Several methods can be implemented such as part number identification or various visual management tools or a combination of both. Many part number identification strategies have been used in the industry, including adding prefixes or suffixes to existing part numbers, or changing part number schemes with specific identifiers detailing compliance, or new dedicated compliant part numbers. Upon deployment of part number identification systems, the material must be marked or labeled with the compliant part number. Material integrity is critical as it is the foundation of the compliance assurance system.

6.4.2 Visual Management Techniques

The second segregation tool is visual management or signs to identify the compliant material and eliminate the opportunity for error. The visual management practices are recommended for the entire manufacturing process including, but not limited to, receiving, stockroom, SMT and through-hole components, wave solder, all applicable rework stations, and field or customer returns area. The entire SMT line should utilize signs to identify that the product and chemistry being used are lead-free. The practice is used for dedicated and non-dedicated equipment to clearly identify and communicate lead-free compliance to all employees and potential customers. The goal is to identify all applicable lead-free equipment such as the SMT process as stated above, hand solder, assembly, wave solder, product repair, inspection, test, and mechanical assembly process.

Given the industry transition to lead-free, product identification of assembled product and material staged for manufacturing becomes critical. The signs segregate the materials and heighten the awareness for all employees, as well as reduce the risk for error and cross-contamination during assembly operations such as hand solder or a manual or equipment rework station.

The last two areas to deploy visual management and control are in indirect materials and field or warranty returns areas. These areas are at a greater risk for cross-contamination, given that leaded and lead-free product and material will be stocked together. *Indirect materials* are defined as lead-free solder pastes, lead-free solder bars, lead-free hand core solder, and equipment and tooling such as solder stencils, labels, adhesives, product repair items and engineering change order items, as well as PCB wires or wires used for product applications. It is recommended to obsolete any leaded or non-RoHS-compliant material such as labels, adhesives, and product repair and wire items to eliminate the opportunity for exposure. As noted in many industry publications, wire products have been confirmed as non-RoHS-compliant given lead in the insulation of the wire which is not exempt and non-RoHS-compliant. The final area of deployment is the field or customer returns area. This area will contain mixed product for a longer duration compared to the manufacturing assembly area, and in this area, lead-free identification of material, equipment, and tools should follow the visual management deployment process noted above.

6.4.3   Due Diligence and XRF Verification

To substantiate compliance, a verification program utilizing XRF technology (X-ray fluorescence analysis) can be implemented. The deployment of such of program will provide a method for the first line of detection while collecting verification data for the compliance assurance system. XRF technology is the phenomenon in which a material is exposed to X-rays of energy, and as the X-ray strikes an atom in the sample, energy is absorbed by the atoms. A core electron is ejected from an atom, if the energy is high enough. To fill the void left behind, a different electron will takes its place. This change in position gives off energy that can be detected by a fluorescence detector. The energy which is released is specific to its element which is then discerned by the XRF analyzer. The XRF technology is available as a handheld and nondestructive test device. Incorporating an XRF verification process requires a structural and financial commitment given the infrastructure required including training, technical analysis, and data recording. The technology is also costly, ranging from $35,000 to $50,000 per unit depending on features and manufacturers. This methodology replicates the expected process of RoHS compliance verification used in the EU per the RoHS enforcement guidance document dated May 2006. The purpose of the XRF verification program is to complement the supplier certificate of compliance and material declarations. XRF verification provides a screening process for the due diligence or compliance assurance system. Figure 6.14 is an example of the handheld XRF technology testing of PCBs.

6.4.4   XRF Testing during the Material Handling Process

The XRF screening program should verify key points of the process including receipt of material, stockroom, and manufacturing process including equipment parameters and manufactured materials. The initial verification step is the receiving and incoming inspection process. The testing of components at this stage verifies compliance with

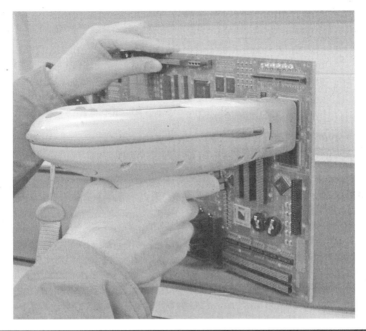

FIGURE 6.14 XRF testing of a printed-circuit board.

the manufacturer's certificate of compliance. All noncompliant materials can be returned to the supplier before product is assembled. Industry sampling standards can be used to determine the amount of testing to be conducted. Pareto charts noted below can be used as a reference which highlights some of the noncompliant commodities and components. During testing, it is important to remember the following critical attributes of the RoHS directive:

- RoHS directive maximum permitted limits for the banned substances: 1000 ppm or 0.1 percent for lead, mercury, hexavalent chromium, polybrominated biphenyls (PBBs), polybrominated diphenyl ethers (PBDEs) and 100 ppm or 0.01 percent for cadmium.

- The definition of homogenous material. *Homogenous material* is a material which cannot be mechanically disjointed into different materials. The term *homogenous material* is also defined as uniform composition throughout. Components may contain multiple homogenous materials; thus testing different areas and parts of the product or component become critical. An example is a semiconductor package or integrated circuit (IC) which includes the molding material, coating on the lead frame, lead frame or termination alloy material, and lastly the bonding wires internal to the package.

- The final attribute is the many exemptions in the RoHS directive. For example, if lead is detected over the allowed 1000 ppm limit, the component may be exempt given the lead-in-solder exemption (known as RoHS-5), the lead-in-alloy exemption for metals, or lead-in-glass exemption for chip components.

Given the attributes noted above, the analysis and subsequent investigation of the XRF testing results become important. Many suppliers do not readily declare their exemptions on their certificate of compliance or material declaration documents. If an XRF reading is detected above the permitted limits, an escalation process is recommended to research the manufacturer's compliance documentation and RoHS directive exemption listing. This process can be deployed not only for the receiving process but also for the stockroom or anywhere material is transacted. One area of deployment may be stock transfers which transfer stock from one part number location to a compliant part number location. Without deploying or verifying the stock during this transfer, the error for cross-contamination of product is greatly increased.

6.4.5   XRF Testing during the Manufacturing Process

The second key detection point is the manufacturing process. The goal is to verify that the process is manufacturing a compliant product. Several verification points should be created to monitor compliance throughout the manufacturing process. The first point, assuming that the specific product includes SMT technology, is during the screen-print operation or during first article inspection of the PCBs. The intent is to verify that the lead-free solder paste has not been cross-contaminated with lead with the verification being completed on each build side of PCB (top and bottom). Verification is required at the solder paste screen-print operation since opportunities for error exist. If the product is found to be contaminated with lead later downstream, the result could be financially catastrophic. The next detection point should focus on other soldering operations such as wave solder, solder pot, and hand solder operations. The routine verification of these processes and resulting assemblies should be incorporated into the control procedures. With the opportunity for cross-contamination at these processes, incorporating statistical process control using XRF verification will provide data supporting compliance.

The record retention of the XRF results becomes the proof and history of verification and supports the compliance assurance system, when the product is questioned by a RoHS enforcement authority. Streamlining the record retention process will decrease the time to respond if a noncompliance is detected.

6.4.6   XRF Verification Findings

An example of an XRF verifications system was the one performed at Benchmark Electronics, Hudson division. All findings that have been confirmed noncompliant by the manufacturers and suppliers have been collected to date. They were categorized with Pareto charts by banned substances and component types. The majority of the tests indicated the presence of lead (29), cadmium (4), hexavalent chromium (2), and PBDE (1). They were found mostly in wire/cable assemblies (14), hardware (5), and connectors (4). Note that XRF technology cannot detect some RoHS banned substances (hexavalent chromium and PBDE) but can detect the base element (chromium, bromine). The hexavalent chromium and PBDE banned substance were confirmed after investigations with the manufacturer. Upon detection of the banned substances, the noncompliant material was purged and segregated. Corrective action requests were issued to the suppliers with compliant material being supplied for the current and subsequent production builds.

In summary, the largest noncompliant banned substance detected was lead, given its prolific use in the electronics industry. The highest noncompliant component type was wires and/or cables which has also been noted in many industry publications and seminars. This occurs when lead is contained within the insulation or PVC of the wire or cable. The Pareto data collected reflect the product mix of a specific manufacturing operation and related exemptions of the finished product. The occurrence will differ among manufacturing operations given various products, suppliers, and declared exemptions.

6.4.7   Hybrid Assembly Process

Given the industry transition to lead-free products and related exemptions, the management of the supply chain has been, and continues to be, challenging. Various commodities have obsoleted tin/lead or non-RoHS-compliant manufacturing part numbers, with only RoHS-compliant parts being offered. This transition has caused a new classification for PCBs, defined as hybrid. The term *hybrid* is defined as a PCB which uses tin/lead solder and contains lead-free plastic ball grid arrays or other types of components. The term can also relate to a lead-free PCB assembly which contains leaded components. A lead-free plastic BGA with SAC305 alloy requires collapse of the balls or spheres which occurs at approximately 217 to 220°C to make a reliable solder joint. A traditional tin/lead reflow profile typically specifies a maximum reflow temperature of 208 to 218°C. The flowchart in Fig. 6.15 provides a guideline for "leaded" hybrid assemblies.

With the introduction of lead-free and hybrid PCB manufacturing, the process control of thermal reflow profiles becomes critical. The process windows of the maximum reflow temperature and maximum

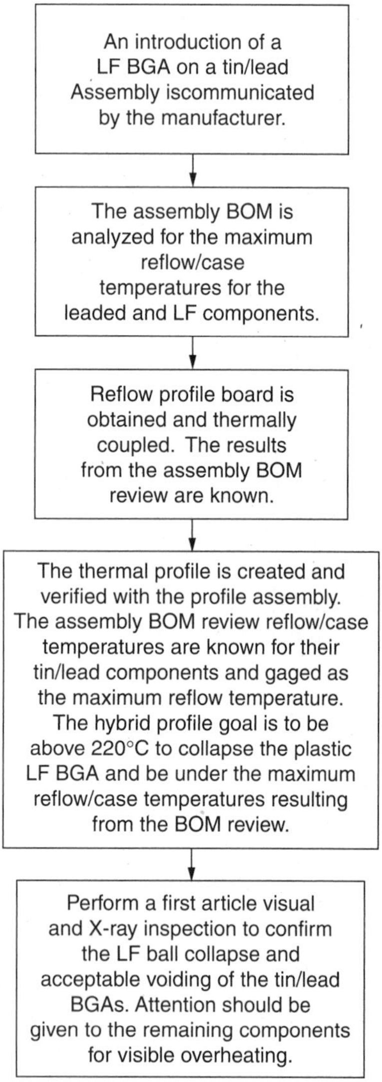

FIGURE 6.15
Hybrid assembly flowchart.

An introduction of a
LF BGA on a tin/lead
Assembly iscommunicated
by the manufacturer.

↓

The assembly BOM is
analyzed for the maximum
reflow/case
temperatures for the
leaded and LF components.

↓

Reflow profile board is
obtained and thermally
coupled. The results
from the assembly BOM
review are known.

↓

The thermal profile is created and
verified with the profile assembly.
The assembly BOM review reflow/case
temperatures are known for their
tin/lead components and gaged as
the maximum reflow temperature.
The hybrid profile goal is to be
above 220°C to collapse the plastic
LF BGA and be under the maximum
reflow/case temperatures resulting
from the BOM review.

↓

Perform a first article visual
and X-ray inspection to confirm
the LF ball collapse and
acceptable voiding of the tin/lead
BGAs. Attention should be
given to the remaining components
for visible overheating.

case temperature of the components are much smaller for hybrid and lead-free assemblies when compared with a tin/lead reflow process (see Fig. 8.4). Given the minimal process window, the verification and control of the lead-free profiles become paramount. This verification should be completed during initial specification and subsequent revisions to confirm that the parameters (maximum temperatures, ramp, TAL, etc.) are acceptable to internal procedures and the solder paste manufacturer's recommendation. Once a profile has been proved, identification and change control must be documented and communicated to all users. Consult Sec. 4.7.2 for more details on the supply chain track and Sec. 7.3.2 for additional details on hybrid assembly solutions.

6.4.8   Hybrid Assembly Case 1: Lead-Free Components Assembled with Tin/Lead Solders

This intermixing was shown to be viable, but with careful control of the process, for most types of lead-free components including BGAs. Cross sections of a lead-free plastic BGA using tin/lead solder paste were performed. They showed 100 percent uniform diffusion of the tin/lead solder with the SAC alloy in the BGA. The black lines contained within the cross sections are the lead mixing which is uniform from the package to the PCB pad. The mixing resulted from a minimum reflow temperature of 225°C. The key result is uniform diffusion and collapse of the lead-free solder balls in the plastic BGA when using tin/lead solder paste.

Alternatives should be deployed before introducing lead-free BGAs into a tin/lead solder process. It is recommended that last-time material buys and conversion of the entire product to lead-free be investigated before the introduction of a lead-free BGA into a hybrid tin/lead assembly process.

Most passive and lead frame components including 1206's, 0805's, 0603's, 0402's, PQFP208's, and SOIC16's that are lead-free (RoHS-compliant) are fully compatible with the tin/lead assembly process provided the tin/lead reflow profile reaches a peak temperature of 210 to 218°C (target of 215°C), the time above 183°C is 60 to 90 s, and the time above 200°C is a minimum of 30 to 35 s. This special reflow profile helps in the diffusion of the tin on the lead-free component frame into the tin/lead solder, thus creating a reliable joint. Some lead-free surface finishes for passive and lead frame components are not compatible with the tin/lead process.

6.4.9   Hybrid Assembly Case 2: Tin/Lead Components Assembled with Lead-Free Solders

The use of components with a tin/lead finish on a lead-free assembly is strongly discouraged because most tin/lead components are rated to 230°C and the lead-free reflow profile will have a peak temperature range of 240 to 248°C. The higher temperatures associated with lead-free would likely destroy or damage many tin/lead components. It might be necessary to assemble the tin/lead components in a separate tin/lead reflow operation after the lead-free components have been assembled. An example of a hybrid assembly is given in Sec. 6.6.2, showing the reflow profiles used for lead-free and tin/lead successive operations.

6.5   Training of the Assembly Associates and Inspectors

Given the impact of lead-free manufacturing from material selection to field returns, training of the manufacturing operation becomes paramount. The lead-free conversion is the biggest change in electronics manufacturing in more than a decade. Having multiple alloys (tin/lead and lead-free) creates the need to have dedicated manufacturing processes or shared processes with detailed process control. The training time to prepare and facilitate will depend on how large the organization is and the length of the content. The legislation and process control training may take several hours to prepare with a 1-h training class being sufficient to facilitate. The visual inspection and technical process training may take up to 2 days to prepare and facilitate the training. The visual inspection and technical process training will need classroom time and on-the-job training (OJT) to successfully deploy the methodologies of lead-free manufacturing.

6.5.1   RoHS Directive Training Requirements

The prerequisite for all technical training should be a review of the RoHS directive as well as others current and pending. The training of the RoHS directive should detail some key points which will provide the background and awareness of the requirements. It is important to explain the scope of the directive and the impact it will have on the organization. The specific key points noted during the training should discuss the impact to electronic equipment, the July 2006 affectivity date, and the banned substances which are identified as lead, mercury, cadmium, hexavalent chromium, polybrominated biphenyls, and polybrominated diphenyl ethers. The maximum permitted levels and definition of homogenous levels should also be highlighted. Providing examples of homogenous levels will provide the audience with an enhanced understanding. The final legislation topic may be the unfortunate discussion of the penalties if a product is found to have a banned substance and the fines, loss of revenue, and even imprisonment, which may be levied if deemed by the RoHS enforcement authorities and a court of law. With the changes affecting all departments within an organization—production, materials, purchasing, engineering, quality, IT systems, training, and finance—the recommendation is to translate the training to the level of the audience by having separate sessions detailing the legislation, impacts, and potential future changes which may need to be investigated. The website http://www.berr.gov.uk/innovation/sustainability/index.html is the United Kingdom's Department for Business Enterprise and Regulatory Reform and is a helpful resource to assist in interpretation of the RoHS/WEEE directives and exemption training.

Another beneficial website is http://www.aeanet.org/Government Affairs/gajl_EnvOverview0306.asp which is the website for a trade group called Advancing the Business of Technology. This group is tracking worldwide environmental legislation including legislation in the United States at the federal, state, and local levels. Some of the information on the website is for members only. The Electronics Industry Alliance website http://www.eia.org/ also tracks environmental legislation.

6.5.2 Banned Substance Training Examples

The next key point should include where the RoHS banned substances occur during the manufacturing process either during PCB manufacturing or mechanical system level manufacturing. A few examples could inform that lead has been used for many years and is included in terminations, solder, BGA spheres, inside the body of various components, and used as an alloy element in metals. It is also a good idea to provide a brief summary and content of the remaining banned substances of the RoHS directive. The list below contains examples of banned substances and where they may be found in electronics manufacturing.

- Cadmium and lead are often used in plastics and cables. One of the common industry findings is lead in the insulation (PVC) of wires and cables.
- Mercury is used in pigments for paints and dyes.
- Hexavalent chromium is common in metallic surface finishes. A metallic surface finish can be on mechanical, SMT PTH components.
- PBBs and PBDEs (flame retardants) are common in plastics and possibly in PCBs.

6.5.3 Visual Inspection Training

The next phase of training should consist of the technical training. One of the initial sessions should be inspection criteria and review of the changes between tin/lead solder and lead-free solder. Given the different alloys and the first alloy change in the industry in years, the retraining of the organization becomes important. The difference between the alloys from an inspection standpoint is primarily the visual appearance of the solder joint. This includes all SMT, wave solder, and hand solder operations and the related solder appearance and connection. Acceptable lead-free joints may exhibit graininess or dull appearance, the wetting contact angles are greater, and the solder does not flow as well as traditional tin/lead solders. As an example, a lead-free solder paste deposit may not cover the entire pad after reflow whereas the tin/lead solder is more likely to cover the pad completely. Refer to Figs. 6.5 and 6.6 which had identical sized paste deposits prior to solder reflow.

6.5.4   Material Process Control Training

The next phase of training must deploy the process control methods of Sec. 6.4. The training of the process control methods permeates many organizations including material receipt, stockroom, and production area. Some of the key points to communicate during this training are material identification and segregation methods. Providing the training foundation for the material receiving and stockroom organizations will empower the team to address and correct many of the cross-contamination opportunities.

6.5.5   Manufacturing Process Control Training

The deployment of process control methods for the production or assembly area is complex, given the multiple operations in electronics manufacturing. The following checklist can be used to simplify the message while providing clear direction. It is directed toward PCB assembly operators:

1. Lead-free signs for SMT equipment, material racks, rework stations, and hand assembly and solder stations

2. Lead-free product traveler signs which follow the manufacturing process and specific product

3. Lead-free dedicated green color-coded tools for the SMT line touch-up including paste spatulas and tweezers

4. Lead-free dedicated green color-coded tools for the rework, hand assembly solder area including solder iron tips, hand cutters, and tweezers

5. Lead-free dedicated green matting on various workbenches

The above checklist will provide awareness to the manufacturing associates and to all who enter the manufacturing area. As stated above, providing the training foundation for the manufacturing organization will empower the team to address and correct many of the cross-contamination and continuous improvement opportunities.

6.5.6   Technical Process Training for Support Organizations

The next training phase should cover the technical organization, including process engineering, quality, and test. The training should discuss an overview of the process control changes and manufacturing impacts and attributes of the lead-free solder. Table 6.5 details the process temperatures and times required from a tin/lead and lead-free soldering process during SMT reflow, wave solder, and rework. In summary, the temperatures required for lead-free are higher, thus requiring many process and product changes including reflow, wave solder, and rework profiles. As stated in Sec. 6.4, the components must be rated to 260°C to withstand the higher temperatures.

	Tin/Lead	Lead-free
Pasty range	183°C	217–221°C
Reflow	210–220°C	240–248°C
TAL	60–90 s	60–90 s
Wave/selective solder	260°C	265–280°C
Wave contact time	3–5 s	5–8 s
Rework (BGA case)	230°C	260°C

TABLE 6.5 Process Temperature and Times Comparison

The above information in Sec. 6.5 details the formalized training and the many organizations affected. With the training of the various departments comes awareness of the impact and resulting challenges and obstacles ahead to achieve the goals of the organization. Many impromptu training and brainstorming sessions will result with peripheral organizations as the lead-free and RoHS transition has a rippling effect throughout.

6.6 Conversion of a High-Volume Product from Tin/Lead to Lead-Free Including Qualification of Through-Hole and SMT Rework

6.6.1 Overview

Benchmark Electronics conducted a lead-free qualification for one of its Massachusetts customers in January 2006. This customer designs and manufactures computer systems based on open industry standards. The lead-free conversion was done on a limited number of an existing tin/lead high-reliability medical products. This product is referred to as the Maverick Board and is 7.5 in by 9.2 in, 16 layers with 7 layers being power or ground, 84 mils thick; and the board was populated with 1694 total components on both sides. The components include both surface mount (e.g., BGAs, SOICs, and discrete components) and through-hole technology (i.e., connectors). The tin/lead version has been in production for an extended period.

The Maverick Board is exempt from the RoHS directive, being a medical product. However, the product is representative of many of the customer's products, and the results would provide a good insight into the viability of lead-free assembly. Moreover, by conducting this study early, the issues and challenges of transitioning to lead-free electronics assembly can be identified and addressed in the most cost-efficient manner.

6.6.2 Summary of the Assembly Process and Results, Using Hybrid Assembly

It was decided to assemble 18 lead-free Maverick Boards evenly divided among the three surface finishes of OSP, ImAg, and ENIG. Each finish has advantages and disadvantages. All flux chemistry was organic acid, and the alloy for all soldering, touch-up, and forced rework was SAC305. The solder paste used on the bottom side was vendor A, and the paste used on the topside was vendor B. These were the two top-performing lead-free organic acid pastes identified during the TV1 work covered earlier in this chapter. The laminate material chosen to meet the higher thermal requirements associated with lead-free assembly was the Isola IS410 product. This product has a Tg of 180°C and is specially formulated for high performance through multiple lead-free thermal excursions. The board component mix includes 15 lead-free BGAs on the bottom side and 11 lead-free BGAs on the topside. A photograph of the topside of the Maverick Board appears in Fig. 6.16.

For components to be acceptable for RoHS-compliant (lead-free) assembly, they need to be rated to 260°C. Of the 1694 components, 1675 were available with RoHS compliance and the required temperature rating. For the remaining 19 components, either they were not available with a RoHS-compliant finish, or it was cost-prohibitive for the vendor to change the component to RoHS compliance for this study. These 19 included 17 BGA components that could not withstand the lead-free temperature profile of 260°C.

The presence of the tin/lead components, including the 17 BGA necessitated a third and fourth lower temperature reflow profile for

FIGURE 6.16 Maverick Board.

their installation on both sides after all lead-free components (including BGAs) were installed to ensure a correctly functioning product. Tin/lead solder paste was micro stenciled on each site prior to the placement and reflow of the tin/lead BGAs.

The initial lead-free profile was a ramp to peak with a ramp of 1 to 2°C/s, peak temperature of 240 to 248°C, and a 60- to 90-s TAL above 217°C. The follow-on tin/lead profile was a similar ramp-to-peak profile with a peak temperature of 200 to 205°C and 60- to 90-s TAL above 183°C. The lower tin/lead peak temperature ensured that none of the lead-free components rereflowed. All profiles were done with an air environment. The solder pot temperature used for through-hole assembly was 285°C, the boards were preheated to a topside temperature of 120°C, and the typical dwell time with the solder was 9 to 11 s.

The lead-free boards were assembled and subjected to visual inspection, ICT, and functional test. The results were compared with historical records for the tin/lead version of the product, and the following conclusions were drawn:

- The mean defect level for the lead-free assembly process is not influenced by the three board surface finishes (OSP, ENIG, or ImAg).

- There was no statistical difference for the mean defects per board between the tin/lead and lead-free process.

6.6.3   Reliability Results from the Thermal Cycling of the Maverick Board

An evaluation of adequate life for the Maverick Board was performed using the Coffin-Manson model to estimate an equivalent use capability from accelerated life testing results. These results were compared to normal use of the product which is one power cycle and a temperature range of 10°C per day. Six boards (two OSP, two ImAg, and two ENIG) were subjected to the following thermal cycling parameters:

- -40 to 85°C temperature extremes (125°C temperature range per cycle)

- Minimum 30-min dwell at each extreme temperature

- Temperature stabilization reached when the board temperature changes less than 2°C in 5 min

- Maximum temperature slope of 20°C/min

- 500 cycles of temperature extremes

- No vibration

- Boards electrically tested prior to the start of thermal cycling and after each 100th cycle

Five boards successfully completed 500 cycles, and one board failed between the 400th and 500th cycle. The applicable data for 400 thermal cycles were entered into the Coffin-Manson model per Table 6.6 using the tin/lead exponent (2.5), since tin/lead and SAC305 solders are both soft metal systems and most studies to date have established comparability in reliability performance between these two alloys in many applications. More details on this model are discussed in Chap. 3. The calculations were performed by Gene Bridgers, a reliability engineer at Mercury Computer, using an Excel-based program that he developed.

The results of the Coffin-Manson model analysis for the Maverick Board are shown in the right portion of Table 6.6. The temperature cycle acceleration factor per test cycle is calculated at 552.4. The equivalent use is 605.4 years per board, or 5,303,300.9 h of equivalent use during the 400 test cycles.

Based on six boards with no failures before the 400th cycle, the normal (life) use during the 400 cycles would be 3632.4 years or 31,819,805.2 h. The point-value mean time between failure (MTBF) for the six boards at the 500th cycle is 31 million h based on 31 million h of equivalent operation with one failure.

The end product company engineers were confident that the reliability of the lead-free version of this product would be equal to or better than the tin/lead version. The tin/lead version of the Maverick Board had not been subjected to thermal cycling which prevented direct comparisons. However, there were a number of tin/lead components including BGAs on the lead-free version, and the thermal cycling data also indicated they were equivalent.

Coffin-Manson Model Model Parameters	Units	Entered Value	Results
Use range temperature	°C	10	Temperature cycle acceleration factor per test cycle 552.4
Test range temperature	°C	125	Equivalent use (years) 605.4 uses years
Coffin-Manson exponent		2.5	
Number use cycles per day		1	Equivalent use (hours)
Number of test cycles		400	5,303,300.9 use hours

TABLE **6.6**   Coffin Manson Calculations for Maverick Board

Moreover, the data indicated there was no difference among the PCB surface finishes.

The Maverick Board that failed thermal cycling between the 400th and 500th cycles was an OSP card. It was subjected to failure analysis, and two modes of failure were determined, but it could not be established which failure mode occurred first. Chapter 3 presents greater discussion of these failure modes.

1. Failure mode 1 plastic BGA at U10, 304 spheres, 1.27-mm pitch: The CTE mismatch between the package and the PCB caused the PCB pads in the corners of the BGA to lift. The continuity was still intact between the sphere and PCB pads, but as more pads lifted, increased stress was placed on the sphere-to-package interface until this interface fractured in several locations.

2. Failure mode 2 ceramic BGA at U9, 360 spheres, 1.00-mm pitch: Similar to failure mode 1, a CTE mismatch placed stress in the corner joints. The sphere-to-PCB pad interface remained intact, but as additional pads lifted, increased stress was placed on a PCB etches emanating from a corner pad. This failure differs from the equivalent type of ceramic BGA failure on a tin/lead assembly in which the stress normally causes a fracture in the bulk solder of the joints. Figure 6.17 illustrates this failure mechanism: the board failed thermal cycling between the 400th and 500th cycles, failure location is ceramic BGA at U9, and the arrow denotes the fractured PCB trace.

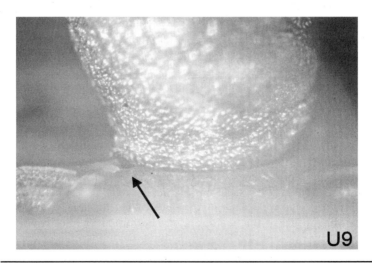

FIGURE **6.17**   Fractured PCB trace in Maverick Board.

6.6.4   Through-Hole Forced Rework on Practice Boards 1 and 2

Practice boards 1 and 2 were selected for development work prior to reworking through-hole components on the Maverick Board. The objective was to identify the optimum method of rework to achieve IPC class 3 through-hole joints with minimal copper dissolution, pad lifting, and thermal degradation of the laminate. Single, double, and triple reworks were evaluated. A *single rework* is defined as the removal and replacement of a through-hole component, whereas a *double rework* is defined as the second removal and replacement of the component at the same site. The board is allowed to cool to room temperature between rework cycles, and a triple rework follows similarly. Double and triple reworks at the same site would be rare in practice, but a successful double rework may be necessary and also provides a higher confidence level in the integrity of a single rework.

Practice boards 1 and 2 were identical and selected because they had an OSP surface which has a significant risk of copper dissolution. Moreover, the boards had 10 sites for the same through-hole component which allowed different techniques to be assessed at various locations on the board. The sites were far enough apart to minimize the thermal impact of rework from adjacent sites. The board was 14.5 in by 16.5 in, 74 mils thick, and 4 of the 10 laminate layers were power or ground.

Practice Board 1: Rework Using Solder Fountain

Sites 1 through 4 on practice board 1 were reworked completely on a lead-free solder (SAC305) fountain with the solder temperature set at 280°C. The original installation of the components was also done on the same solder fountain. All boards were preheated to a topside temperature of 150°C and underwent a single or double rework. The rework nozzle style was varied to provide a "stagnant" or "full flow." Stagnant flow is defined as the solder making intimate contact with the bottom of the board but not flowing over the sides of the nozzle, whereas *full flow* is defined as the solder making intimate contact with the bottom of the board and flowing over the sides of the nozzle. The objective of the study was to determine if the stagnant flow reduced the copper dissolution as compared with the full flow. The *contact time* for the rework is defined as the time the solder is making intimate contact with the bottom of the board. The flux used was vendor A organic acid flux paste.

Measurements of the copper thickness on the PCB on sites 1 through 4 were taken as shown in Fig. 6.18. The top layer pad copper is on the topside of the board, and the bottom layer trace copper is covered by masking. Both locations have negligible risk of copper dissolution and can be considered baselines for the original copper thickness.

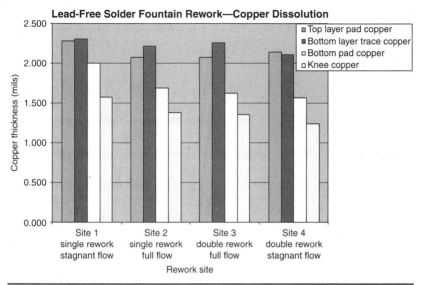

FIGURE 6.18   Copper thickness measurements for sites 1 through 4 on practice board 1 (OSP).

The following are some observations from the practice board 1 and 2 rework results:

- All four sites were acceptable. There was no evidence of excessively thin or fractured knees, nor was there any significant via wall thinning near the surface. Copper thickness was above the IPC class 3 minimum of 0.001 in (1 mil).

- Copper dissolution increased for the double rework as compared with the single rework, which was expected, but the increase was smaller than anticipated.

- The difference in copper dissolution between the stagnant and full flow was not significant, which was not anticipated.

- By the end of the double rework process, the bottom side pads have lost approximately one-quarter of their original copper, regardless of the method used.

Practice Board 2: Rework Using Forced Convection

Practice board 2 was used to assess a different rework process in which forced convection was used to remove the connector and remove solder from the through holes. The solder fountain was only used to install the component. This process significantly reduced the contact time on the solder fountain, and it was hypothesized this would minimize copper dissolution. One connector location was subjected to a double rework, and another connector was subjected to a triple rework; and the two sites are referred to as 2X and 3X, respectively.

The parameters for the hot air component removal and cleaning of the holes were as follows:

- Board topside preheat for component removal and cleaning of the holes was equal to 160°C.
- Peak temperature of the component plastic body was equal to 245°C.
- Peak temperature of the solder in the barrel was in the range of 225 to 235°C.

The parameters for the component installation on the solder fountain were as follows:

- Board topside preheat = 150°C
- Temperature of the solder = 280°C
- Contact time = 8 to 9 s

Copper thickness on the PCB was measured, and the data appear in Table 6.7. The top layer pad copper is on the top of the board, and the bottom pad copper is covered by masking and have negligible risk of copper dissolution and can be considered baselines for the original copper thickness on the PCB.

The measured results were acceptable for the 2X and 3X reworks per IPC class 3 criteria. The copper thickness at the knees did not go below 1 mil, and the amounts of copper dissolution on the 2X and 3X rework were equivalent, which was surprising since it was anticipated the dissolution would increase on the 3X rework as compared with the 2X. The final copper thickness on the bottom pad copper and knee copper was greater on practice board 1 than on practice board 2, which was not anticipated. However, based on the copper thickness of the top layer pad copper and bottom layer trace copper, it appears that practice board 1 had more copper prior to the start of rework than practice board 2, and this would explain the results. The cross

	2X	3X
Top layer pad copper	1.872	1.764
Bottom layer trace copper	N/A	2.103
Bottom pad copper	1.357	1.364
Knee copper	1.001	1.035

TABLE **6.7** Copper Thickness for the 2X and 3X Sites on Practice Board 2 (OSP)

sections of the 2X and 3X reworks were acceptable per IPC class 3 criteria. The amount of copper dissolution was acceptable, and the small amount of pad lifting is consistent with tin/lead through-hole reworks. The flow-through of solder to the topside was 100 percent, there was minimal voiding, the wetting was good, and there was no thermal degradation of the laminate.

6.6.5 Maverick Board Through-Hole Rework

The Maverick Board is 84 mils thick, 16 layers with 7 ground and power planes. Compared to practice boards 1 and 2, it represents a significantly more difficult rework challenge in terms of thermal load. The component selected for rework on the Maverick Board was the 16-pin header at location J6. All cross sections showed that the rework met all specifications at 2X and 3X sites.

Maverick Board 3: OSP Surface Finish, Double Rework on the Solder Fountain

The results from practice boards 1 and 2 indicated a double rework done completely in the solder fountain would be successful. It was decided to employ this process on Maverick Board 3, which was an OSP surface finish. This process is faster than the alternative forced convection process, because it takes less time to remove a component on a solder fountain than on a forced convection platform.

The solder fountain parameters for the rework of the 16-pin header at location J6 on Maverick Board 3 are as follows:

- Topside board preheat temperature = 150°C
- Temperature of the SAC305 solder = 280°C
- Contact time for each rework = 18 to 20 s
- Total contact time for both reworks = 36 to 40 s
- Flux = vendor B flux

The *contact time* for each rework is defined as the total time required for removing a component and installing a new one while the solder is making continuous contact with the bottom of the board.

The double rework was not acceptable in a 16-pin header on Maverick Board 3 after double rework on the solder fountain, as illustrated in Fig. 6.19. The top view at left shows a pad lifting, the interior view in the middle shows a hole wall pulling away, and a bottom view at the right shows a fractured knee. The fractured knee is caused by copper dissolution due to excessive contact with the SAC305 molten solder. This is a known risk with lead-free SAC305 rework and has been documented previously. This board would be scrapped which could have significant cost and delivery implications. Note that the fractured knee could not be detected with visual inspection

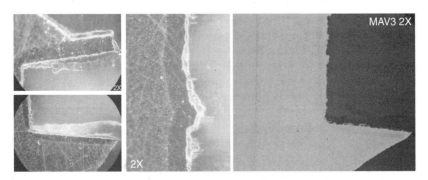

FIGURE **6.19** Maverick Board failures after double rework on the solder fountain.

and required destructive cross sections for detection. The board also exhibited through-hole pull-away which is not acceptable and also exhibited undesirable pad lifting.

The Maverick Board is more thermally massive than practice boards 1 and 2 and required longer contact times on the solder fountain which likely caused the unacceptable copper dissolution. A different condition of Maverick Board OSP finish rework was later successfully used with forced convection, shown in Maverick Board 9.

Maverick Board 4: ImAg Surface Finish, Single Rework Using Forced Convection with the Solder Fountain

Based on the unacceptable results for the double rework of Maverick Board 3 on the solder fountain, it was decided to perform a single rework incorporating the forced convection platform for component removal and through-hole cleaning in conjunction with the solder fountain. The solder fountain was only used to install the new component, which significantly reduced contact time on the fountain which in turn minimized the risk of copper dissolution. The initial forced convection parameters were identical to those of practice board #2, but the plastic body of the 16-pin header melted and it was determined that the plastic was only rated to 220°C which made it difficult to remove the component. The topside preheat of the board was increased from 160 to 180°C, the peak temperature on the plastic component body was lowered to 220 to 230°C, and the component was removed from the board just before the plastic body melted. The temperature of the solder in the through-holes was between 220 and 225°C. The solder fountain was used to install the new component with the following parameters:

- Topside board preheat temperature = 150°C
- Temperature of the SAC305 solder = 280°C
- Contact time for installation = 8 to 9 s

The single rework was successful in minimizing copper dissolution but induced some PCB thermal degradation as manifested by pad lifting and hole pull-away. The pad lifting varied from none to a maximum of 2 mils. The hole pull-away was less than 10 percent along the entire length. The pad lifting and hole pull-away are acceptable but not preferred. It was believed the thermal degradation was most likely caused by the high topside preheat (180°C) combined with the extended time at this temperature required to reflow the component, remove the component, and vacuum the holes. The higher preheat temperature was required to minimize the melting of the plastic body (rated at 220°C) of the through-hole component, and the preheat would have been lowered if the through-hole component were rated to a higher temperature. A contributing factor was likely the four reflow cycles required during the original assembly which corresponded to bottom side lead-free, topside lead-free, bottom side tin/lead, and top side tin/lead.

Maverick Board 9: OSP Surface Finish, Double Rework with Forced Convection in the Solder Fountain

Based on the generally positive results with Maverick Board 4, it was decided to conduct a double rework using the same technique on an OSP Maverick Card. The forced convection and solder fountain parameters were identical to those of Maverick Board 4, and the total contact time on the solder fountain was 16 to 18 s (two installations).

This process significantly reduced copper dissolution as compared with the process used for Maverick Board 3. The solder fill was greater than 75 percent as required for IPC-610 class 3 criteria, wetting was good, and the voiding was minimal. However, slight pad lifting was detected, and this is likely related to the high board preheat temperature required to compensate for the low temperature rating of the 16-pin header. Moreover, this board was subjected to a double BGA rework prior to the through-hole rework, and the board was reflowed four times during the original assembly. The increased number of thermal cycles was likely a factor in the thermal degradation. Standard lead-free boards are normally reflowed twice during the original assembly, and it is extremely unlikely four reworks would be performed on the same board.

Maverick Board 8: ENIG Surface Finish, Double Rework with Forced Convection in the Solder Fountain

The double rework technique used for Maverick Board 9 was used for Maverick Board 8, which is an ENIG board. Cross section of joints shows that the process meets IPC class 3 criteria, but there is evidence of slight pad lifting which is not preferred. The thermal degradation was likely caused by similar factors as for Maverick Board 9. Copper dissolution on Maverick Board 8 was not an issue, which was expected given the surface finish was ENIG.

6.6.6 BGA Rework

All Maverick Boards were preheated to a topside PCB laminate temperature of 160°C prior to the BGA being reflowed. The applicable reflow parameters for the BGAs were as follows:

- Maximum case temperature = 260°C
- BGA joint temperature = 240 to 245°C
- TAL 217°C = 75 to 90 s

Double rework was conducted on a 304-pin lead-free plastic BGA located at U29 on the topside or a 280-pin lead-free BGA located on the bottom side at U510. The rework was completed on each of the three board surface finishes, and vendor A was used for solder paste and liquid flux. Flux paste was used to clean the site after the BGA was removed, and solder paste was microstenciled on the site prior to installation of the new BGA. A double rework on the same site is rare, but a successful double rework may be necessary; and it provides a higher confidence level in the single rework.

Maverick Board 5: OSP Surface Finish, 304-Pin BGA

The 304-pin BGA at U29 was subjected to a double rework on Maverick Board 5 (Mav5), which means the original BGA was removed and replaced and the process was repeated after the board was allowed to cool to room temperature. The surface finish of the board was OSP, and vendor A's solder paste was used. The following are the applicable rework thermal parameters:

- Topside board preheat temperature = 160°C
- Maximum case (BGA package) temperature = 259°C
- Peak joint temperature = 245 and 243°C
- TAL 217°C for the joints = 82 and 85 s

The board was tested electrically before and after the double rework and passed both tests. Cross sections of the board showed results which are acceptable per IPC class 3 criteria. There is good wetting, minimal voiding, good symmetry, and minimal copper dissolution on the pad. Moreover, no damage was observed to the PCB. However, the press pin connectors on the edge of the board were thermally degraded and had to be replaced after the double rework to complete the electrical testing. The press pin components were not rated to 260°C.

Maverick Board 10: ENIG and ImAg Surface Finishes, 304-Pin BGA

A double rework was performed in a similar fashion on the 304-pin BGA on Maverick Board 10 which had an ENIG and ImAg surface finish. There was no evidence of thermal degradation, and the BGA joints met IPC class 3 criteria.

6.6.7   Maverick and Practice Boards 1 and 2—Conclusions

- The forced convection platform in conjunction with the solder fountain dramatically reduced the copper dissolution observed on the Maverick Board as compared with rework on the solder fountain only. The former is the recommended rework process for thermally massive boards such as the Maverick Card.

- The reliability of the lead-free version of the Maverick Board was deemed acceptable per the calculations made using the Coffin-Manson model.

- Overall, the lead-free through-hole rework processes evaluated were successful. The solder fountain or a combination of the hot air platform and the solder fountain is a viable alternative, and the selection of the optimum rework process is predicated on the design and construction of the board.

- Additional optimization is required for the through-hole rework process to minimize pad lifting.

- Double rework of BGAs on an ENIG, ImAg, and OSP surface finish was acceptable, and there was no evidence of thermal degradation to the PCB or BGA.

- All components should be rated for the higher temperatures associated with lead-free assembly and rework. The through-hole header reworked on the Maverick Board was only rated to 220°C and melted during hot air removal. Similarly, the press-fit connectors, which were exposed to the bottom heaters during through-hole and BGA rework, exhibited thermal degradation. A 260°C temperature rating is recommended for all lead-free components.

6.6.8   Forced Rework of a Through-Hole Component on the Alpha Board

The alpha board was selected for additional forced through-hole rework because the board was larger and heavier than the Maverick Board. The alpha board was 18 in by 19 in, 0.103 in thick, with 16 layers including 6 layers of 2.6-mil-thick copper. The surface finish was ImAg, and the assembly weighed approximately 10 lb when fully populated. The component selected for rework was the 240-pin DIMM connector which was approximately 5 in long. There were a total of 32 connectors on the board.

The first rework on the alpha board was attempted completely on the solder fountain. The board was preheated to a topside laminate temperature of 125°C, and the SAC305 solder temperature was 285°C. The rework nozzle was sized slightly larger than the connector, and after 35 s of contact time on the nozzle, the connector joints were just starting to liquefy; but the process was stopped due to excessive contact time on the nozzle. The board was inspected, and 60 percent of the bottom side pads were partially or fully dissolved, making it evident this mode of rework was not acceptable given the fact that significantly greater contact time would have been required if the component had been removed and replaced. The board would have to be scrapped which would result in significant cost and delivery implications.

The method of rework introduced on the Maverick Board that incorporated forced convection to remove the component and vacuum the through-holes was implemented on the alpha board. It reduced the contact time on the solder fountain from greater than 35 s to 7 s because the fountain was only used to install the component. The results were successful with minimal copper dissolution, minimal thermal degradation to the board, and good solder flow to side 1.

A double rework was attempted on a separate site on the alpha board, but it was not successful due to excessive copper dissolution. One of the failures is shown in Fig. 6.20, revealing unacceptable copper dissolution at the solder joint knee.

The total contact time on the solder fountain for the double rework on the alpha board was 25 s, and the breakdown is as follows:

FIGURE 6.20   Alpha board, second rework—unacceptable copper dissolution at the knee.

Event	Contact Time on the Solder Fountain (s)
Original installation of the DIMM connector	11
Installation of the DIMM connector for the first rework	7
Installation of the DIMM connector for the second (double) rework	7

The total contact time of 25 s for the assembly and double rework of alpha board is comparable to the total contact time of 27 s for the assembly and double rework of the Maverick Board. However, the double rework on the Maverick Board was successful whereas the double rework on the alpha board was not. Observations made during the reworks indicated the dissolution always began on the smaller side of the bottom side capture pad, which tended to lift up off the board. Once it lifted, the pad was exposed to SAC305 on both sides, which increased the temperature of the pad and the rate of copper dissolution. The root cause of the pad lifting appeared to be insufficient radial contact distance. As shown in Fig. 6.21, the radial

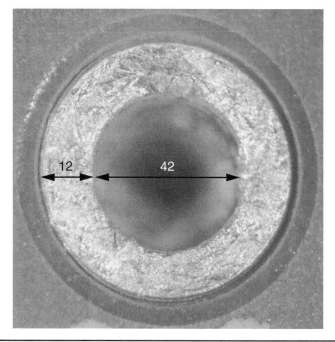

FIGURE **6.21** Maverick card, bottom side pad for the 16-pin header showing recommneded 12-mil radial contact distance.

contact distance for the Maverick Board was 12 mils whereas it was only 6.6 mils for the alpha board. To minimize the risk of copper dissolution during assembly and rework, it is recommended the radial contact area be similar to the Maverick Card or approximately 12 mils. This change is also recommended for the alpha board although, if space becomes an issue, it is possible a 10-mil radial contact area may be sufficient.

Based on the above work, the Maverick product is qualified to be converted to lead-free, assuming the supply chain issues of component availability and temperature ratings for selected components are resolved.

6.7　Costs of Lead-Free Manufacturing

Over the last several years, electronic manufacturing companies have been under pressure to cut manufacturing costs due to the weak economic demand and outsourcing trend to low-cost labor regions. The transition to lead-free manufacturing is requiring companies to assess their cost structures as many are increasing the cost of business and shippable products. The following is a sample of the many areas which are impacted by the lead-free transition, creating increased costs for electronic manufacturing.

6.7.1　Resources and Equipment Costs

The lead-free conversion is requiring companies to research and develop test vehicles and conduct reliability studies qualifying the manufacturing process. These projects require resources to manage, scope, design, manufacture, inspect, and analyze the test vehicles. Material and tooling must be purchased to manufacture the test vehicles with extensive quality and reliability testing. In parallel, the qualification of equipment must be completed to prove compatibility to new product introduction. Upon completion of the product and process qualification, capital equipment may have to be modified or purchased to implement lead-free processes. Some examples of required capital equipment may be wave solder or solder fountain machines compatible with and dedicated for the lead-free products. Reflow ovens and rework stations will also have to be investigated to ensure they can meet the higher-temperature demands. Additional costs will be required for process control and visual management as detailed in Sec. 6.4. Examples of process control are XRF equipment, dedicated tooling, and visual management tools such as signs, matting, or other items to identify lead-free processes.

6.7.2　Training and Process Modification Costs

The next cost which impacts an organization transitioning to lead-free is training as stated in Sec. 6.6 and accompanied by process modifications and learning curve impact for lead-free manufacturing. Training the organization and process modification can be time-consuming, especially during the initial stages of lead-free conversion. The development and modification of processes to accommodate lead-free will involve many organizations and processes. Some examples are as follows:

- Increased need to control moisture-sensitive materials
- Process control during reflow, wave solder, and rework profile development
- Part introduction and engineering change notification (ECN)
- Due diligence verification via compliance testing (XRF technology or third-party laboratory)
- Incoming quality assurance and receiving process modification
- Supplier risk management and audits for compliance
- Record management and retention for due diligence

6.7.3 Material Procurement and Process-Related Costs

The material costs for direct components have varied by commodity and manufacturer. Beginning in 2005, the cost for compliant or lead-free material was higher than for the noncompliant (leaded) version. Since the RoHS deadline, this cost has switched for most commodities (passives, actives), with the noncompliant components becoming more expensive, and some are obsoleted. The PCB costs have increased due to the new laminate materials and higher-temperature requirements. Some compliant components may not be available within the manufacturing lead time, resulting in lot changes and termination modification charges. Termination modification is the process of reballing or retinning the component terminations to become compliant, resulting in increased part delivery time due to this added process step. An example is an integrated circuit, which may be retinned by removing the tin/lead component finish and then adding a compliant termination finish. During this process, the maximum case temperature during the assembly should be considered if the lead frame is coated in the body of the component. Retinning a component does not ultimately determine compliance; it may require analysis of the homogenous materials. Component availability may also increase process manufacturing costs. Noncompliant components can exhibit lower maximum case temperatures and thus cannot be processed through a lead-free solder reflow process. This will require additional processing steps and ultimately added costs; this case is defined as a hybrid assembly and was discussed in Sec. 6.4.7.

6.7.4 Infrastructure and Summary Costs

An additional cost area to monitor is the electricity costs which will increase given the higher temperatures required for lead-free alloys. As an example, studies have been completed stating that the average electricity cost for the lead-free reflow process will increase by 15 to 20 percent given the higher temperatures required. Last is the cost associated with archiving traceability records and possible infrastructure setup of this process in the event a product is challenged by the EU or customer. Records must be maintained to substantiate the compliance assurance system and provide a defense of the product, if requested. Table 6.8 provides a summary of the cost impacts of lead-free manufacturing. All the cost data in Table 6.8 were derived by the authors and based on supplier surveys or best estimates.

Cost Area	Cost Impact	Transition Cost Impact by %	Sustaining Cost Impact by %
Material procurement	Cost increase	5 to 30 (depending on commodity)	0 to 15 (depending on commodity)
Lead free solder & process material	Cost increase	30 to 70 (depending on material)	10 to 30 (depending on material)
Material receipt, inspection, and inventory	Cost increase	5 to 15	5 to 10
SMT and assembly process (build & modification)	Cost increase	10 to 20	5 to 10
Test process (ICT, functional)	No cost impact	No cost impact	No cost impact
Rework and inspection	Cost increase	10 to 20	5 to 10
IT systems for traceability, data retention	Cost increase	20 to 50 (depending on system and process implemented)	10 to 20 (depending on system and process implemented)

TABLE **6.8** Cost Impact Summary

The above summary provides an estimate of costs. The cost impact may be greater in some organizations, given many factors such as fluctuating metal prices, material transition costs, and the company's specific processes and infrastructure.

6.8   WEEE, Green Recycling, and Beyond

Manufacturers will be required to improve the design of their products to avoid the generation of waste and to facilitate the recovery and disposal mandated by WEEE as follows:

1. The substitution of hazardous substances such as lead, mercury, cadmium, hexavalent chromium, and certain brominated flame retardants

2. Measures to facilitate identification and reuse of components and materials, particularly plastics

3. Addition of traceability and environmental labeling

4. Measures to promote the use of recycled materials such as plastics in new products

Within the EU, recycling centers and consortia have been established to provide all recycling services to companies that do not have the resources and infrastructure. Given the infrastructure creation and sustaining costs, joining the recycling consortia may be the obvious choice, even for larger companies.

The deployment of the WEEE and RoHS directives is stimulating creativity in the next steps of green electronics. This creativity is being designed into products, resulting in new environmentally friendly materials being implemented and new technologies being created. In the PCB industry, products are being designed without bromine-based flame retardants or halogen-free product. Various companies are looking ahead at the next round of banned substances which can be linked to the impact of the chemical restriction directive also known as REACH in the EU. This environmental trend toward banning substances is requiring companies to be knowledgeable and ready for global environmental legislation. See Chap. 5 for typical global product development.

Acknowledgments

The chapter authors would like to thank Richard Russo and Gene Bridgers of Mercury Computer Systems; Greg Morose of the Toxic Use Reduction Institute; and Steve Beck, Paul Bodmer, David Cavanaugh, Richard Garnick, Dan Gibbs, John Goulet, Mike Hasty, Mike Havener, Jon King, Dave Roy, Kim Sharpe, Bruce Tostevin, and the entire corporate lead-free team of Benchmark Electronics for their assistance and contributions.

References

Farrell, R., S. Mazur, C. Morose and R. Russo. "Transition to Lead-free Electronics—Case Study." *IPC/JEDEC International on Lead-free Electronic Components and Assemblies*, Montreal, Canada, August 2006.

Farrell, R., et al. "Transition to Lead-free Electronics Assembly Case Study— Part II — Product Reliability and Forced Rework." *IPC/JEDEC Global Conference on Lead-free Reliability and Reliability Testing for RoHS Lead-free Electronics*, Boston, Mass., April 2007.

Gadnick, R. "TV3 Reliant Lead-free Assembly Final Report." Internal Benchmark Electronics report, March 2006.

Gibson, G., and P. Schaffner. "Functional Test Implementation Specification." Benchmark Electronics Internal Document 07-07726-00, Winona, Minn.

Hamilton, c., m. Kelly, and P Snugovsky. "A Study of COPPER DISSOLUTION during Lead-free PTH Rework Using a Thermally Massive Test Vehicle." *Circuits Assembly*, May 2007.

"Lead-free, High Temperature Fatigue Resistant Solder; Final Report of NCMS." Project 170503-96034, July 2000.

Ma, L., et al. "Reliability Challenges of Lead-free (LEAD-FREE) Plated-Through-Hole (PTH) Minipot Rework at Intel Corporation." *IPC/JEDEC Global Conference on Lead-free Reliability and Reliability Testing for RoHS Lead-free Electronics*, Boston, Mass., April 2007.

Morose, C. "New England Lead-free Electronics Consortium – Phase 3 Efforts." *9th Annual IPC/JEDEC Lead-free Conference*, Boston, Mass., December 2005.

"Round Robin Testing and Analysis of Lead-free Solder Pastes with Alloys of Tin Silver and Copper; Final Report." IPC Solder Products Value Council, A Research Report by the Lead-free Technical Subcommittee, July 2005.

"Originally distributed at the SMTA Successful Lead-Free/RoHS Strategies Conference." Boxborough, MA, June 20–21, 2007.

CHAPTER 7

Establishing a Master Plan for Implementing the Use of Green Materials and Processes in New Products (Product Development Going "Green")

Richard A. Anderson, Ph.D.

Tyco Electronics Wireless Systems Segment (M/A-COM, Inc.), Lowell, Massachusetts

7.1 Introduction

In this chapter, the basic principles of new green electronics product design are discussed and how they impact and are impacted by design for the environment (DfE). To succeed, the design team must understand how the properties and application of environmentally friendly materials and processes affect the product. The product development team must plan and select these items based on not only an environmental focus but also all the design criteria. For these reasons, the environmental engineer should be an integral part of the design process.

7.2   An Overview of Green Electronics Design Basics

The design issues for green electronics products are outlined below.

7.2.1   Functionality

The most obvious design rule is that the product should function and perform properly. The most successful materials and processes used in electronics should be selected to achieve performance. The team must consider whether the environmentally friendly materials replacement could alter the performance. If so, can the design parameters be changed so that these materials can be employed successfully? Will the processing conditions affect the performance of any of the components used to create the electronic product? Therefore the following must be understood:

1. *Conductivity.* Will the materials and processes affect speed, input power, noise, or electrical shielding? Fortunately, copper, the mainstay of electronics and electrical equipment, is not banned and has been used on circuit boards almost exclusively as the conductor of choice. Copper, however, does tarnish by oxidation, reducing conductivity significantly.

   Most circuit boards use a solderability preservative coating as a finish to prevent oxidation which also inhibits soldering efficiency. For years, the finish of choice was hot air solder leveling (HASL) where the completed circuit board is fluxed and then dipped in molten solder that wets the copper surface and renders it easily solderable using mild, nonacidic fluxes. The molten solder is immediately stripped off using a high-pressure hot air knife to remove all but the thinnest coating. In areas not requiring solder, solder mask generally protects the copper from tarnish.

   The new lead-free solders require additional heating to temperatures 30 to 40°C higher than tin/lead based solders, making HASL less practical and more likely to cause thermal damage to the board. Fortunately, alternatives exist including the following:

   • *Organic solderability preservatives (OSPs).* These organic coatings vaporize or lift off in molten solder during the soldering process. These are quite inexpensive and effective but tend to have short shelf lives.

- *Immersion tin.* This is a tin plating that is also inexpensive and easily soldered to but may grow tin whiskers or tin pest. Tin finishes can be applied improperly, making soldering difficult.

- *Immersion silver.* Now this is a fairly mature process but more expensive. Immersion silver can tarnish with time if not stored properly. Unlike earlier silver finishes, the new immersion silver is unlikely to grow dendrites via electromigration, resulting in short circuits.

- *Electroless nickel overplated with immersion gold (ENIG).* ENIG is quite stable with time under most storage conditions, but the cost is related to gold pricing. At high frequencies, the nickel under the ultrathin gold (4 to 12 μin) can reduce RF performance. To combat this, thicker gold can be used but is expensive and can embrittle solder joints, both with lead-free and with lead-bearing solders. The design team should note with caution that electroless nickel plating can also contain lead and cadmium, although new plating baths not containing either have been developed.

- *Lead-free HASL.* Lead-free HASL that uses tin-copper or other lead-free alloys has been developed. Lead-free HASL requires higher-temperature processing with the increased possibility of board damage.

- *Other, newer, alternative finishes.* These may include electroless nickel-immersion palladium-immersion gold (ENiPiG), a lead-free finish often used on package leads; electroless nickel-autocatalytic gold (ENAG); and direct immersion gold (DIG).

Solderbility finishes on component leads must also be lead-free. The most common and inexpensive finish is 100 percent matte tin which can grow whiskers under certain conditions. ENiPiG discussed above is also a suitable lead finish (Romm, Lange, and Abbott, 2001).

2. *Resistivity.* Will changes in the resistance of the surface finishes create electrical noise, cause loss of information, output signal, and/or overheating?

Resistivity is the inverse of conductivity, but a few additional points should be made. Of all the common metals, silver has the highest conductivity with copper and gold not far behind. Solder, while quite conductive, has a higher resistivity. The soldering process essentially is the alloying of tin with a nonmelting metal, forming an intermetallic compound which freezes at soldering temperatures. Intermetallic compounds (IMCs) are ordered metal structures (tin-copper-tin-copper-tin) that behave more as compounds than as metals with higher resistivity and other nonmetal properties such as brittleness. Since the new lead-free solder alloys are mainly tin, these IMCs can form a continuous layer with undesirable properties such as high resistance.

3. *Dielectric properties.* Will changes in these properties create frequency shifts, phase shifts, loss of power, coupling, or result in breakdown? At high (RF) frequencies, current is carried in the electromagnetic field surrounding the conductor, not in it. The new soldering processes often require the use of different materials capable of withstanding the higher temperatures and longer soldering time required. These materials can have different dielectric properties that can alter the performance of a circuit, especially at high frequencies. Increased dielectric loss can absorb energy and affect performance. Different dielectric constants will alter 50-Ω line impedance values, affecting the circuit adversely. Changing materials therefore may require an extensive redesign to ensure performance is not affected. Since additive brominated flame retardants such as PBBs and PBDEs must be replaced, the new additives can also alter dielectric properties even if the base materials are not changed. See Fig. 7.1, showing the effect of using a new polymer with a lower dielectric constant and a 50-Ω line width using circuit analysis software "tx line calculator," supplied by Applied Wave Research, Inc., 2210A Graham, Redondo Beach, California. The electronic circuit must be redesigned.

4. *Corrosion resistance.* Will the product still have the required lifetime and not undergo catastrophic failure while operating in the field environment? One of the banned materials is

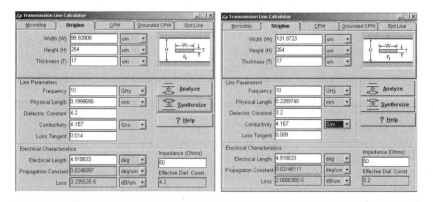

Figure 7.1   Effect of using a new polymer with a lower dielectric constant as a 50-Ω line width. (Courtesy of Applied Wave Research.)

hexavalent chromium. This finish goes by many names but is essentially a coating that inhibits corrosion very effectively on a number of different metal surfaces. Many electronic housings, cabinets, and cases used hexavalent chromium for corrosion protection. It is often detected by a yellowish tint on bolts and surfaces, for example. There is no drop-in replacement for all hexavalent chromium applications, and many replacements effective on one surface are ineffective on others. Trivalent chromium, permanganates, and other replacements exist; all work well in some applications but not in others.

5. *Appearance.* Do the environmental substitute materials and processes make the product unsightly, or feel "funny" during handling? Appearance and feel are rather subjective concepts, but if the replacements affect these qualities, product desirability, and therefore sales, may suffer.

6. *Strength.* Will the product fail in use, withstand accidental drops or impacts, or weaken with fatigue over time? The replacement solders do not have the same mechanical properties as tin/lead. They are stronger and do not relax (lower stresses by stretching as easily) and are not as resistant to high strain rates and some cyclical stresses. In addition, their higher processing temperatures can impart higher thermal mismatch stresses between components and the circuit board. These effects can put additional stresses on solder joints, which may fail in shorter time periods than might occur in tin/lead. The failure mechanisms for lead-free solder are also different than for tin/lead in many cases. Reliability testing must be repeated for circuits employing the new materials. In some cases, reliability tests shown effective for tin/lead may not apply to the new solders, and false lifetime data may result. This is not to indicate that the new solders are not as reliable, only that the reliability has not yet been effectively established for all conditions and applications.

7.2.2   Cost, Size, and Weight

These issues are of prime concern to the design team, and size and weight are increasing in importance, especially with portable electronic devices. The issues facing the design team are as follows:

1. *Redesign.* As noted above, the use of new materials may require a circuit redesign. This option is expensive but may be required for the continued success of the product.

2. *Raw materials costs.* Initially, replacement materials will cost more since the supply is low and processing has not been optimized. Supply and demand may also cause cost increases.

3. *Processing costs.* Due to the higher temperatures and longer times required, energy costs and throughput may suffer and additional maintenance may be needed to combat the wear and tear introduced by higher temperatures. For example, some of the hexavalent chromium replacement coatings take longer to apply and require more steps, increasing labor charges. Some of these costs may be mitigated by lower disposal costs associated with green materials.

4. *New equipment.* Higher processing temperatures may also require new equipment designed to withstand these temperatures. In addition, new processes may require new equipment because existing equipment cannot handle or apply the new materials correctly.

5. *Qualifying new vendors.* While many vendors have developed lead-free soldering processes, smaller-volume suppliers are still in short supply since they simply will not have the workforce required to develop lead-free and other assembly processes until demand requires it. Therefore, one may have to audit and work with suppliers to ensure their processes are robust and effective. Failure to do so may result in latent (hidden) defects that do not become apparent until the product is in use by customers. This may require costly replacements and even recalls if safety is an issue.

6. *Additional weight of replacement "environmentally safe" materials.* This is important in automotive, aerospace, and portable products and must be considered.

7. *Development of new product.* Sometimes it may be more effective to develop a new product than redesign for the above reasons because any changes may jeopardize the product.

7.2.3 Manufacturability

You cannot wait months for a new process if you need the new product now. Processes will have to be developed concurrently with new, compliant products to ensure success. Cost, discussed above, is only one of the issues. Yield and throughput also affect costs but can affect schedules and deliveries as well. Availability is an issue that affects all parts of the chain and must be included in design for manufacturability. Existing noncompliant parts simply cannot be used in green electronics.

1. *What processes have to change?* Earlier, lead-free soldering processes were discussed. Any compliant product has to employ the higher-temperature processes to produce good products. Qualifying processes, whether in-house or at a vendor, becomes critical. There are major drawbacks to the new soldering processes,

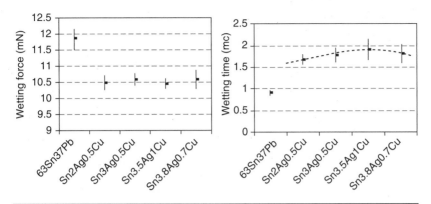

FIGURE 7.2   Differences in wetting force and wetting time between lead-free and tin/lead solders.

such as that the wetting time and wetting forces for lead-free solders are longer and lower (Anderson et al., 2003; Crowe and Feinberg, 2000) so processes must be altered to avoid things like tombstoning and other defects resulting in expensive, more difficult rework. See Fig. 7.2 for data on differences in wetting times and forces, presented by the author at the Annual New England IMAPS Symposium in May 2003 at Boxborough, Massachusetts.

2. *Time and temperature for soldering.* Can the materials selected withstand the increased stresses associated with lead-free reflow, wave soldering, and especially rework? Rework is a major issue because in a localized process the higher melting temperature increases thermal gradients, which can induce damage at or around the rework site. Often the entire circuit must be preheated to minimize these gradients and reduce dwell time at the rework site.

3. *Proper equipment.* This was discussed above.

4. *Inspection.* As noted elsewhere in this book, compared to tin/lead, lead-free solder joints have a significantly different appearance. Inspection must detect the following:

   • Is the dull lead-free solder joint a good connection or a "cold" solder joint? See Fig. 7.3, which is a comparison between a tin/lead and lead-free BGA solder bumps, showing that both are good, though with different appearances. They were manufactured for the TURI TV-2 DoE.

   • Due to higher lead-free wetting angle, is this a good joint or a cold solder-joint? The lead-free solder joint wetting angles in Fig. 7.4 shows the dramatic effect the atmosphere

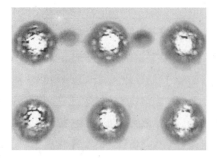 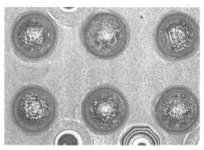

FIGURE 7.3 Comparison between a tin/lead and lead-free BGA solder bumps.

(air versus nitrogen) can have on joint wetting angle and appearance. In the figure there are solder joints produced using the same conditions, solder pastes, and parts, but the joint on the left was reflowed in air while the one on the right was reflowed under nitrogen.

- Since lead-free solder shrinks more on solidifying than tin/lead, is there fillet lifting or an actual joint failure? An example is shown in Fig. 7.5 where, after the solder wets the copper pad, it shrank during cooling, from the top pulling the molten solder up off the pad as it solidified. This condition is not a failure, although in inspection it looks like a crack.

- Have the higher temperatures resulted in any rework or assembly damage? IPC has developed new guidelines for lead-free assembly, and new training associated with these guidelines is in place. These are found in IPC-A-610D referred in the reference section.

- Are the processes "in-place"? By "in-place" one must ensure the yields are high and the processes do not induce any hidden damage. FMEA (failure mode effects analysis) should be performed to ensure yields and reliability are

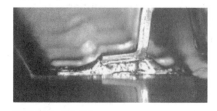 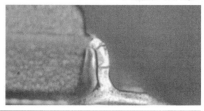

FIGURE 7.4   Lead-free solder joint wetting angles.

Figure 7.5 Fillet lifting due to solder shrinkage. (From Syed, 2001.)

high. For example, proper storage of moisture-sensitive parts for reflow should be available. The higher reflow temperatures and longer times at these temperatures can induce damage by rapid release of absorbed moisture if parts are not stored properly. Parts used in tin/lead reflow that had a moisture sensitivity level (MSL) of 1 (on special storage required) may be down-rated to 3 (exposure to ambient for more than one week could result in moisture release damage). If so, they must be sealed against moisture ingress or stored in a dry box until ready for assembly; otherwise they will have to be baked to slowly drive out the absorbed moisture before they can be reflow soldered.

a. Process windows. A *process window* is defined as the ranges over which the processing conditions (time, temperature, atmosphere, solder paste, part size, shape, and weight) can be varied and still obtain a good, reliable assembly. These ranges and their interactions should be determined and documented to ensure a high yield. Tools to develop these windows can be found in references on statistics and Six Sigma processing conditions such as Shina (2002) and Crowe and Feinberg (2000).

b. Process statistics (C_p, C_{pk}). Variable C_p is a measure of process capability or how repeatable your process is, while C_{pk} is the process capability index or how scattered the results are within the process window C_p. A C_{pk} of 1.33 (4σ) is considered a robust repeatable process. Process control charts and measurements are required to establish and maintain a good process.

7.2.4   Reliability

What effects do the new materials and processes have on the following?

- The new lead-free soldering processes mainly require temperatures roughly 25 to 35°C higher than those used in tin/lead processing. As a result, higher temperatures can and do induce damage in electronics. This is particularly true in many polymers or plastics now being exposed to higher assembly temperatures, but can also arise due to thermal gradients and thermal expansion mismatch. In addition to thermally induced effects, interactions with the environment have to be considered where new materials are employed under conditions for which they may not have yet been tested. This is especially true for the hexavalent chromium replacements where, like the new solders, they are replacing materials that have been in use for years in numerous environments. Since in many cases the damage is not immediately detectable, it can result in operational failures. This latent damage thus may reduce the lifetime of the circuit. As was stated earlier, FMEA should be employed to assess and then investigate potential failures. Types of failure modes are as follows.

- Thermal mismatch coupled with the higher yield strength of lead-free solders can increase stresses on solder joints. Higher melting points and thermal stress relief temperatures coupled with increased solder joint rigidity may prevent solder joint relief, resulting in higher stresses. Greater solder joint shrinkage upon solidification can add to these stresses. Failure may occur due to creep (constant stress over long periods) or fatigue (cyclical stresses occurring in on/off operation). Parts with markedly different thermal expansion coefficients than that of the circuit board such as ceramic packages, ball grid arrays, capacitors, and resistors can be particularly susceptible. High-temperature operation lifetime (HTOL) and power cycling reliability testing may be required to assess these effects.

- Thermal degradation of polymers resulting from elevated processing temperatures must also be considered. In some cases these can exacerbate thermal mismatch stresses, weakening the circuit board to solder pad copper adhesion (bond strength) and enhancing stresses on the thin copper foils used to create the conductive traces. Loss of adhesion and mismatch can also affect plated blind, buried, and through vias interconnecting multilayer boards, causing cracked via copper that results in intermittent or permanent open-circuit conditions. Boards should be tested for these failures using interconnect stress testing (IST) or even thermal shock to establish whether these stresses can induce failures. Similar to the loss of adhesion to copper, polymer degradation can cause loss of adhesion between the glass weave and epoxy or other laminate resin. This can allow moisture ingress along the glass fibers. The presence of moisture between oppositely biased internal traces or vias can allow electromigration of copper from the positive to the negatively biased interconnect, resulting in a copper filament short circuit over time. These are called *conductive anodic filaments* (CAFs). See the filament growth shown in Fig. 7.6. Testing under operation in high humidity and elevated temperature (called CAF testing) may be required unless the projected use environment precludes humidity effects. Many of the new lead-free compatible laminate materials are listed as IST and CAF resistant. Board materials should have a thermal degradation temperature above about 320°C to ensure reliability.

- The lack of a lead-rich phase in the solder joint can create a tin-to-copper interface, allowing a continuous copper/tin intermetallic compound to grow. This growth is accelerated by temperature and is more likely to occur initially during elevated-temperature assembly. These intermetallic layers can have high resistance and can be brittle, causing performance issues and mechanical failures over time. Copper/tin IMCs under 100 percent tin (and even under tin-rich solders and plating) can increase the tendency for tin whiskers to grow, which may cause short circuits over time. The use of a nickel barrier plating between tin and copper can reduce whisker growth tendencies by preventing copper/tin IMC formation both on component

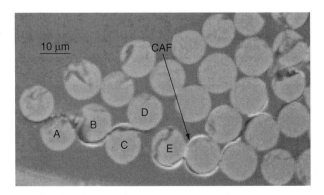

FIGURE 7.6 Conductive anodic filament (CAF) growths.

leads with 100 percent tin solderability plating and on circuit board traces.

- Hexavalent chromium coatings are used in electronics for several reasons. Corrosion prevention is a major application, and hexavalent chromium adheres well to and protects many metal surfaces including steel, zinc diecast, and aluminum. One desirable property of hexavalent chromium is scratch protection. Hexavalent chromium has been shown to migrate and "fill in" scratches to prevent corrosion in these areas almost as effectively as it does on pristine surfaces. Surface migration in hexavalent chromium replacements has not been verified so salt fog and other corrosion resistance testing should include samples with scratches in the coating that penetrate to the base metal underneath. A second major use of hexavalent chromium coatings is as a conductive coating. Many metals oxidize and electrical contact between mating housing parts can be jeopardized. Some of the replacement coatings, if applied correctly, are also quite conductive, but the stability of this conductivity should be assessed in relation to the environment in which they are used. A final application of hexavalent chromium is as an improved interface for paint and powder coating adhesion. Adhesion to replacement coatings should be tested via a Tabor abrasion test or the tape test, both before and after salt fog and humidity testing including adhesion around scratches.

- Are the failure modes the same as those for the material(s) being replaced? In many cases they are not the same failure mechanisms. As noted above, lead-free solders are much more susceptible to high strain rates, and any acceleration factors established for tin/lead solders employing high strain rates may not be valid unless such high rates occur in actual use. A good discussion of this related to SMT packaging can be found in Syed (2001). Failures associated with high solidification shrinkage are more important than for tin/lead, and this should be carefully assessed. Elevated-temperature electronics should be carefully tested due to the increased degradation and thermal stresses induced in lead-free soldering. Finally, corrosion, scratch protection, and paint adhesion may be different as noted above.

- Are the failure mechanisms for the new materials even "understood"? In many cases the failure mechanisms are well documented, in others they are not. One must be careful to consider that the effects of testing designed for tin/lead and hexavalent chromium reliability do not induce failure modes in replacements that will not occur in actual use. These modes may be invalid unless the relationship to reliability and lifetime is understood.

- Do the new processes adversely affect the materials that are not replaced? As noted above, the increased stresses and different failure mechanisms can all affect materials not replaced such as ceramic packages and discrete components. Thermal damage to laminates not replaced with more resistant materials is another example. So, often the answer is yes—you have to do your homework.

7.2.5   Meeting Design Requirements

New electronics products have to meet all types of requirements from customers and the marketing department, as well as regulations from any corner of the globe.

1. *Customer specifications.* Requirements may preclude any changes without requalifying the electronics performance and reliability. This is often the case with automotive and military aerospace electronics. As a result, changes to existing products have to be negotiated with customers to ensure their specifications are met. If not, redesign and/or new products may be required where compliance with green initiatives is needed. Consortia and technical societies or organizations such as iNEMI, IPC, the (Massachusetts) Toxics Use Reduction Institute (TURI), and the New England Lead-free Electronics Consortium are helping to understand these effects, recommend procedures to minimize them, and establish good processing methods so their findings should be considered in any new lead-free and/or green product revisions, introductions, or replacements.

2. *Government/agency specifications.* These may mandate no changes and, in some cases, have established a list of "prohibited materials" in some or all electronics. Reasons vary but include the following:

   a. *Safety.* The replacement materials have not been tested for applications where lives are threatened by electronics failures—these are often exempted for green initiatives.

   b. *High reliability and long lifetime.* NASA, military aerospace, and other organizations have developed lists of prohibited materials that cannot be used in some applications. These include banning the use of 100 percent tin finishes and tin/silver/copper solders and may list others. Satellite and other electronics failures due to tin whisker formation have been documented, and in other cases the reliability of replacements is not yet known. To better understand and identify effective replacements, various groups have been created to work on these issues such as JGPP (Joint Groups on Pollution Prevention) and AIA-AMC-GEIA (Aerospace

Industries Association–Avionics Maintenance Conference–Government Engineering and Information Technology Association) LEAP-WG (Lead-free Electronics in Aerospace Project—Working Group). These groups are working on toxics replacements, and a check of their websites for updated information should be made periodically.

c. *Hazardous production issues.* In some cases, replacement of materials involves new processes whose safeties have not been established. In Massachusetts, TURI and other groups are assessing these procedures to ensure the replacement "cure" is not as bad as or worse to workers and environments and the materials and processes being replaced. In most cases this is not true, but alternatives may have to be developed in the few cases that do arise.

d. *Waste stream during manufacturing and for discarded products.* A number of initiatives and legislations have been established for protecting the environment from by-products and discarded electronics containing toxics such as WEEE. Processes must be monitored to ensure any by-products containing these materials are properly disposed of and may require reporting of amounts of lead and volatile organics being discarded. Costs may adversely affect profits in either case due to fines, reporting costs, and waste stream cleanup and should also be considered as a result. Legislation and requirements can vary by country, state, or region and locality.

7.3   Green Product Conversion

To compete in markets where materials bans are mandated, such as lead-free, existing products must be either converted to lead-free or replaced by others. The questions arise, When does conversion to lead-free make sense? When should an entirely new product replacement be developed? To answer these, several items must be identified.

7.3.1   Performance and Features

A noncompliant product should not be converted if it is near the end of its effective salable lifetime and potential replacement new products can offer improvements in capability, size, weight, and operation. On the other hand, old products with desirable features might best simply be made compliant without additional feature improvements. One successful conversion was the green Sony Walkman introduced in 2001. The only significant difference between this product and others was that it was billed as green, prompting environmentally conscious individuals to purchase it over other similar units and thereby gaining a significant market share.

7.3.2   Product Use

If a product will be used in a number of applications or subsystems, it may make sense to offer two versions; one is noncompliant while the other contains none of the prohibited materials cited above. In this case, it will be necessary to effectively separate the two items to ensure cross-contamination does not occur. At Tyco Electronics Wireless Systems Segment—M/A-COM, Inc., this has been done successfully for a number of product offerings using the following practices:

1. Lead-free products have been given entirely new part numbers to distinguish them from those containing tin/lead solder and finishes.

2. Product packaging and literature distinctly identifies lead-free and/or RoHS compliance for all such products.

3. Cross-contamination teams are in place to set up lead-free assembly stations, develop lead-free assembly practices, separate lead-free and non-lead-free materials being used in the production of such items by incoming and finished parts and materials binning, and audit and improve these activities.

4. Purchasing requisition forms have mandatory fields to be filled in that identify requirements for lead-free or non-lead-free parts to be obtained.

5. Methods are developed for removing prohibited materials and refinishing part leads with tin/lead for high-reliability applications where parts containing prohibited materials are the only parts available for exempt and specified applications. A cautionary note here, as initial parts suppliers' warrantees may be voided via this practice. Reballing BGAs is one refinishing operation where this may arise.

6. At Tyco Electronics, as is the case for many other suppliers of high-reliability, "prohibited materials exclusion" parts, the company will continue to supply tin/lead in such applications where the requirements are mandated.

7. Replacement solders such as tin/silver/copper (SAC) solders do *not* necessarily always mechanically fail via the same mechanisms as tin/lead solder. The reliability tests for solder joints were all designed with tin/lead failure modes in mind. Therefore, these tests do not accurately assess the reliability of the new solders since the failures either are predicted by different acceleration factors generated (Shina, 2002) or do not occur in actual operation. In-depth references,

studies, and discussions can be found at such websites as those of GEIA-AIA, iNEMI, ITRI, and others.

8. So what is the reliability of the new solder? Probably very good, but *no one knows!*

9. Qualifying materials, processes, and suppliers are used. Issues such as solder joint voiding, appearance, and shrinkage are dependent on solder composition, solder paste, and reflow/wave solder process. Care must be taken to ensure items such as these are controlled and do not fall outside specified maximum values.

10. The cost-effective replacement for tin/lead finishes is 100 percent tin. But 100 percent tin will grow whiskers. Additives have been shown to reduce this whisker growth tendency, but only if properly applied and treated.

7.3.3   Conversion to Hexavalent Chromium–Free Products

The advantages of hexavalent chromium coatings (hex-chrome) are many and include these:

- Large base of qualified, highly skilled application suppliers.
- Very low cost.
- Fast processing—simply clean properly and immerse in a properly maintained bath or spray, brush on.
- Proven reliability with years of testing experience.
- Conductive finish.
- Excellent finish for paint, powder, and other finishes.
- Self-healing—migrates into scratches and reseals surface.
- Adherent to most metallic surfaces.

Disadvantages of hex-chrome in electronics are mainly associated with safety and the environment and include these:

- *Toxicity.* Exposure limits to workers applying hex-chrome have been and are being lowered, requiring increased worker protection and tighter controls on exposure by handling to end users.
- *Waste stream.* Depleted and unused chemicals require specific controls for disposal. and disposal costs are rising.
- *Reporting costs.* For both disposal and usage, the costs of OSHA/NIOSH reports are increasing as they require more information.
- *Testing for presence.* Most banned materials require testing only for the presence of the specified element: lead, mercury, cadmium. Bromine is only banned in certain specified

compounds. Only the hexavalent form of chromium, not other forms, is banned. Therefore, a test that indicates the presence of chromium may have to be followed with further, expensive testing for the hexavalent form.

A number of replacement coatings that have not been banned were developed. Alternatives to hex-chrome include

- Trivalent chromium compounds.
- Permanganate compounds.
- Organic films, of which there are many different types including paints and powders.
- Other inorganic films such as oxides and other compounds of Al, Ti, Mo, W, Co, Ce, Zr, Si, etc. All produce corrosion-resistant coatings.
- Organometallics, which are quite expensive.
- Combinations of coatings.

While there is no single drop-in replacement for hex-chrome, advantages of hex-chrome replacement coatings are many:

- Some of the trivalent replacement coating solutions are quite safe and could be discarded without hazard to the environment.
- Reporting may not be necessary on replacements depending on solutions and materials used.
- Many are acceptably conductive.
- Most are stable in many conditions. However, this is dependent on the application and the base material to which they are applied.

A large number of replacement coatings have been developed, and selection is based on a number of criteria including these:

- Application methods
- Cost
- Whether the coating has been tested in the environment for which it is destined
- Appearance
- Pre- and postapplication treatments required
- Finding and qualifying vendors—availability

Coating availability and data to support use are spotty so new studies may be required to ensure performance and reliability. Replacements have been shown to be very effective in many applications while the same coatings on different surfaces or in different environments simply are not effective. In addition, some coatings must be applied to different surfaces using different processes and pretreatments to be effective.

7.4 Specifications for Lead-Free and Hexavalent Chromium–Free Alternatives

Specifications for coatings often refer to those listed below. Most automotive and many other companies, defense and federal, state and local governments refer to these or similar specifications. In addition, many may require the use of specific chemicals and processes or that they be selected from an included list. Therefore, these standards must be studied to ensure they are met.

- MIL-883, "Electronics Reliability Standard"
- MIL-C-5541, "Chemical Films for Aluminum and Aluminum Alloys"
- ASTM B921-02, "Standard Specification for Non-hexavalent Chromium Conversion Coatings on Aluminum and Aluminum Alloys"
- ASTM B117, "Practice for Operating Salt Spray (Fog) Apparatus"
- ASTM B602, "Test Method for Attribute Sampling of Metallic and Inorganic Coatings"
- ASTM D1730, "Practices for Preparation of Aluminum and Aluminum-Alloy Surfaces for Painting"
- ASTM D3359, "Test Methods for Measuring Adhesion by Tape Test"
- Federal Std. No. 141, "Paints, Varnish, Lacquer, and Related Materials; Methods of Inspection"
- IEC TC 111 WG 3: CD IEC 111/24/CD, Project 62321 Procedures for the Determination of Levels of Six Regulated Substances (Lead, Mercury, Hexavalent Chromium, Polybrominated Biphenyls, Polybrominated Biphenyl Ethers) in Electrotechnical Products"
- IPC 1065, "Materials Declaration Handbook"
- ISO 2409, "Paint and Varnishes—Cross-Cut Test"
- ISO 3768, "Metallic Coatings—Neutral Salt Spray Test (NSS Test)"
- ISO 4519, "Electrodeposited Metallic Coatings and Related Finishes—Sampling Procedures for Inspection by Attributes"

7.5   Examples of Adverse Consequences to New Green Products Design

7.5.1   Example of Moisture Effect Costs

Moisture is absorbed by plastics, and the moisture release rate is very temperature-sensitive. The higher soldering temperatures associated with lead-free assembly can and will cause moisture to evaporate more rapidly. The release of this moisture may result in die attach cracking and failure or worse, rupture of the plastic encapsulating the semiconductor chip. This is called "popcorn" failure. An example is shown in Fig. 7.7, courtesy of Intel Corporation.

The degree to which a part is susceptible to damage via rapid moisture release depends on the amount of moisture that can diffuse into the materials incorporated in the part and how long it has been exposed to a humid environment before it is used in assembly. This is called the moisture sensitivity level. Table 7.1 (J-STD 020D) shows the levels of sensitivity versus exposure.

The industry has found that the new reflow processes have resulted in a partial MSL increase by one or two levels in many cases. J-STD 033B.1 (Joint IPC/JEDEC Standard for Handling, Packing, Shipping and Use of Moisture/Reflow Sensitive Surface Mount Devices) should be used to properly handle parts with MSL levels higher than 1 (1 means no special handling needed). While catastrophic failures can be found by visual inspection, many failures due to moisture release are hidden and must be detected by scanning acoustic microscopy (SAM). This analysis should be done according to J-STD 035 (Joint IPC/JEDEC Standard for Acoustic Microscopy for Nonhermetic Encapsulated Electronic Components).

Many parts that were MSL 1 for lead-tin processing are now rated MSL 3 for lead-free reflow. As a result, Tyco Electronics Wireless Systems Segment (M/A-COM, Inc.) and other electronics parts suppliers

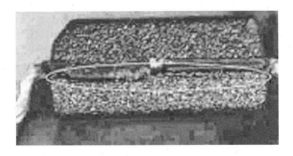

FIGURE 7.7   Crack in package created by SMT reflow moisture release. (Photo courtesy of Intel Corp.)

MSL	Floor Life	
	Time	Condition
1	Unlimited	≤ 30°C / 85% RH
2	1 Year	≤ 30°C / 60% RH
2a	4 Weeks	≤ 30°C / 60% RH
3	168 h (1 week)	≤ 30°C / 60% RH
4	72 h (3 days)	≤ 30°C / 60% RH
5	48 h	≤ 30°C / 60% RH
5a	24 h	≤ 30°C / 60% RH
6	Time on pkg. label	≤ 30°C / 60% RH

TABLE **7.1**   Moisture Sensitivity Levels (from J-STD-020D)

now prebake and seal many plastic packaged ICs in moisture-proof packaging. These extra steps and packaging requirements obviously add to the cost of the parts.

Once the sealed parts are opened, they must be assembled in the allotted exposure time, or they have to be either repackaged or stored in a moisture-free environment such as a nitrogen-purged dry box. Failure to do so will require parts to be rebaked before they can be used. Table 7.2 from J-STD-033B contains these bake-out requirements. As seen, they can be quite lengthy depending on MSL values and conditions. This is more costly than precautionary storage in most cases.

7.5.2   Example of Corrosion and Hexavalent Chromium Replacement with TCP

Tyco Electronics—M/A-COM, Inc. requires effective replacements for hex-chrome coatings be employed except where specified for environmental protection and compliance. Many housings, chasses, and enclosures are made of aluminum. The U.S. Navy published standards for trivalent chromium coating (TCP) and TCP application processes. When properly applied to aluminum, this process meets or exceeds many military and ASTM specifications. TCP is environmentally safe. The U.S. Navy has licensed TCP processes to commercial manufacturers.

To assess TCP, M/A-COM contacted several TCP licensed suppliers. Metalast International, Inc., produces Metalast TCP–HF (hexavalent-free). They coated coupons of aluminum and zinc-plated steel for ASTM salt spray testing at M/A-COM. Tests results showed some surfaces survived over 280 h. To identify local sources, aluminum test parts with threaded holes were supplied, coated with TCP, and then salt spray tested. These parts showed some degree of corrosion, as seen

Package Body	Level	Bake @ 125°C		Bake @ 90°C ≤5% RH		Bake @ 40°C ≤5% RH	
		Exceeding Floor Life by >72 h	Exceeding Floor Life by ≤72 h	Exceeding Floor Life by >72 h	Exceeding Floor Life by ≤72 h	Exceeding Floor Life by >72 h	Exceeding Floor Life by ≤72 h
Thickness ≤1.4 mm	2	5 h	3 h	17 h	11 h	8 days	5 days
	2a	7 h	5 h	23 h	13 h	9 days	7 days
	3	9 h	7 h	33 h	23 h	13 days	9 days
	4	11 h	7 h	37 h	23 h	15 days	9 days
	5	12 h	7 h	41 h	24 h	17 days	10 days
	5a	16 h	10 h	54 h	24 h	22 days	10 days
Thickness >1.4 mm ≤2.0 mm	2	18 h	15 h	63 h	2 days	25 days	20 days
	2a	21 h	16 h	3 days	2 days	29 days	22 days
	3	27 h	17 h	4 days	2 days	37 days	23 days
	4	34 h	20 h	5 days	3 days	47 days	28 days
	5	40 h	25 h	6 days	4 days	57 days	35 days
	5a	48 h	40 h	8 days	6 days	79 days	56 days

TABLE 7.2 Bake-out Requirements (from J-STD 033B.10 Standard)

Package Body	Level	Bake @ 125°C		Bake @ 90°C ≤5% RH		Bake @ 40°C ≤5% RH	
		Exceeding Floor Life by >72 h	Exceeding Floor Life by ≤72 h	Exceeding Floor Life by >72 h	Exceeding Floor Life by ≤72 h	Exceeding Floor Life by >72 h	Exceeding Floor Life by ≤72 h
Thickness >2.0 mm ≤4.5 mm	2	48 h	48 h	10 days	7 days	79 days	67 days
	2a	48 h	48 h	10 days	7 days	79 days	67 days
	3	48 h	48 h	10 days	8 days	79 days	67 days
	4	48 h	48 h	10 days	10 days	79 days	67 days
	5	48 h	48 h	10 days	10 days	79 days	67 days
	5a	48 h	48 h	10 days	10 days	79 days	67 days
BGA package >17 × 17 mm or any stacked die package	2–6	96 h	As above per pkg thickness & moisture level	N/A	As above per pkg thickness & moisture level	N/A	As above per pkg thickness & moisture level

TABLE 7.2 (Continued)

FIGURE 7.8   Salt spray corrosion of improperly coated aluminum test samples and housings.

in Fig. 7.8, showing salt spray corrosion of improperly coated aluminum test samples and housings. Note the white corrosion patches.

Since initial results were very acceptable, M/A-COM, Metalast, and Coating Systems, Inc., of Lowell ran DoE based on precleaning alternatives, bath parameters, and coating times. DoE analyses were used to recommend changes to improve the process. As a result, a fully qualified process with SPC was set up at Coating Systems, Inc. Subsequent testing results of salt spray shown in Fig. 7.9 showed no evidence of corrosion at 168 h under ASTM B117-03 conditions. Images in Fig. 7.9 are of much higher magnification than normal, showing no evidence of corrosion—only slight staining.

7.6   Planning and Implementation of Green Wiring and Cabling

One place lead unexpectedly appears is as a uv or heat stabilizer in polymers. If the lead content of any separable part of the cable or wiring contains more than 0.1 percent lead by weight, it is not compliant. Cadmium is also used as a stabilizer, and polymers containing cadmium cannot contain more that 0.01 percent if they are to be considered green or compliant. Without these stabilizers, polymers such as PVC begin to degrade at temperatures as low as 160°C during the extrusion process. In many cases, the lead- and cadmium-bearing stabilizers are being replaced with zinc, calcium, barium, or organic tin compounds. While these materials are RoHS-compliant, they may impart different characteristics to

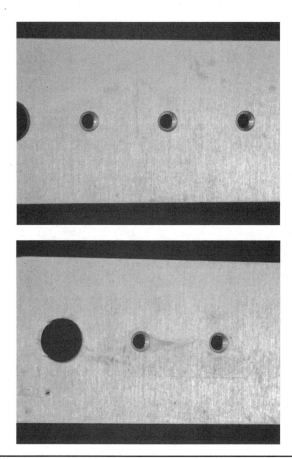

FIGURE 7.9   Results of 168-h salt spray testing of properly coa aluminum parts.

the polymer including changes in RF performance. Exemptions have been sought and exist for some applications.

Pigments used to color-code wire insulation may also contain lead or cadmium. These include but are not limited to red, orange, and yellow hues. These have been replaced with compliant alternatives in many cases, but users should ensure they are not present before claiming RoHS compliance.

7.7   Green Requirements Exemptions

In cases where safety and reliability factors are critical and replacement materials are unproved and in cases where no suitable replacement materials have been found, exemptions are being granted. On example cited above is the reliability of lead-free solders in some harsh applications including military and aerospace use. It is beyond the scope of this discussion to present and justify all these exemptions, but one should be aware of them and where they are applicable. For some exemptions, there are expiration dates. These are fixed based on promising evidence that replacements will be forthcoming and may vary from those granted by the EU to those in China and other countries. It is wise to keep abreast of these exemptions, expiration dates, and new legislation and to plan for replacements or to seek new exemptions where it makes sense.

7.8   Summary

1. Green electronics are viable and can be produced for most commercial and consumer applications. The above discussion identifies materials and processing differences that should be planned for in order to correctly use them as replacements for compliance to green electronics legislation. Requirements can and do vary from country to country, but these differences can be overcome with diligence and planning.

2. Green electronics processes and assembly can be robust and produce reliable, high-performance products for many of the applications where they are required. Research should be done to narrow the process development. A number of studies such as those at the New England Lead-Free Consortium have identified and tested materials and processes capable of replacing noncompliant materials with minimal impact on processing times, costs, and yields.

3. Exemptions exist for sound, valid reasons. As noted above, awareness of where the exemption is applicable and for how long is critical to the success of the product. Specifications may and do prohibit the use of some compliant materials for these exempt uses. Therefore, decisions to supply parts for these applications must be made, and the means for ensuring cross-contamination between exempt and compliant applications must be developed.

4. Products must be tested to prove and ensure their compliance, using recommended means according to current standards. In some instances, new test methodologies must be developed to meet standards. Testing laboratories, in many cases, have this capability at a price.

Acknowledgments

The chapter author would like to acknowledge Stephen Hanlon, Dr. Robert Hilty, Dr. Alec Feinberg, Jama Mohamed, Helena Pasquito, John Penica, and George Wilkish of Tyco Electronics for numerous discussions and efforts associated with development of "green electronics" and design for the environment.

References

Romm, D., Lange, B., and Abbott, D. "Evaluation of Nickel/Palladium/Gold-Finished Surface-Mount Integrated Circuits" Texas Instruments Application Report, 2001.

Anderson, R., et al. "TCP Replaces Hexavalent Chromium Coatings in Enclosures —A Success Story." *Green Supply Line White Paper*, April 2006.

Anderson, Richard, et al. "Massachusetts Lead-free Consortium Status Update." *Annual New England IMAPS Symposium*, Boxborough, Mass., May 2003.

Crowe, D., and A. Feinberg. *Design for Reliability*. New York: CRC Press, 2000.

Huang, B., and N. Lee. "Conquer Tombstoning in Lead-free Soldering." IPC Printed Circuits Expo, SMEMA Council APEX Designers Summit, February 2004.

Shina, S. *Six Sigma for Electronics Design and Manufacturing*. New York: McGraw-Hill, 2002.

Syed, A. "Reliability of Lead-free Solder Connections for Area-Array Packages." APEX—IPC SMEMA Council, Anaheim, Calif., February 2001.

Turbini, L. "Soldering Material Choices: Processes and Reliability Considerations." *CMAP Workshop on Reliability*, Toronto, Canada, June 2001.

Willis, B. "Lead-free Fillet Lifting—It Happens in Reflow and Wave Soldering." *Smart Group White Paper*, published in www.smartgroup.org.

Standards and Websites for Originating Organizations

SAE J1739, "Potential SAE J1739 (R) Potential Failure Mode and Effects Analysis in Design (Design FMEA), Potential Failure Mode and Effects Analysis in Manufacturing and Assembly Processes (Process FMEA), and Potential Failure Mode and Effects Analysis for Machinery (Machinery FMEA)." SAE International, 2002.

IPC-A-610D, "Acceptability of Electronic Assemblies", IPC Standard, February 2005.

J-STD-020-D, "Joint IPC/JEDEC Standard for Moisture/Reflow Sensitivity Classification for Nonhermetic Solid State Surface Mount Devices." August 2007.

J-STD 033B.1, "Joint IPC/JEDEC Standard for Handling, Packing, Shipping and Use of Moisture/Reflow Sensitive Surface Mount Devices." January 2007.

J-STD 035, "Joint IPC/JEDEC Standard for Acoustic Microscopy for Nonhermetic Encapsulated Electronic Components." April 1999.

ASTM B117-03, "Standard Practice for Operating Salt Spray (Fog) Apparatus." ASTM International.

http://www.aia-aerospace.org/

http://www.inemi.org/

http://www.itri.co.uk/

CHAPTER 8

Fabrication of Green Printed Wiring Boards

Michael Taylor

Dynamics Details Incorporated, Longmont, Colorado

8.1 Introduction

Acknowledgment and control of hazardous materials have permeated the design of electronic circuits and circuit boards since their conception: addition of brominated flame retardants to substrates prevented fire propagation from failed circuits; leaded solders made robust joints interconnecting boards and devices; surface finishes helped to make connections reliable after many years of use. Later these solutions were found environmentally unacceptable. While the industry has worked to minimized these materials, by incorporating other less toxic ones, some of the safety solutions themselves have been found or suspected to have their own risks to the environment and health of the population. Incorporation of green laminates into electronics is just one stage of a continuing tradition of risk management in design.

In this chapter, some of the environmental issues facing modern printed-wiring board manufacture are presented in order to share how they are being managed.

8.1.1 Environmental Issues Driving PWB Manufacture

Modern industrial nations faced the issue of environmental contamination and began remedial activity to cleanse damage from prior years during the last half of the 20th century. In July 1970, the EPA was formed in the United States. Since then, the public has supported increased efforts to prevent ground, water, and air contamination from processes producing electronics products and from the products themselves. Most recently, the European Union (EU) passed far-reaching environmental laws affecting the electronics industry known as RoHS and WEEE. Japan, China, Korea, and many other countries have followed suit with laws attempting to regulate products produced or imported into their countries that may contaminate their environment. These regulations, while locally implemented, have global impacts. In a world economy, products are manufactured to the most severe standards around, so that there are no barriers to sales anywhere. As a result, the electronics industry has been impacted in order to accommodate these requirements.

8.1.2 Green Regulations in North America

There are myriad regulations in North America, including individual states, and Canada. More regulations also are constantly being debated in legislation. Figure 8.1 is summary of these regulations, courtesy of Dynamic Details Incorporated (DDi), of Anaheim, California. The most severe of these regulations will affect manufactured products in not only individual states in the United States, but also any country that wants to export to that state.

8.2 Impact of Assembly Processes

To some board fabricators, a requirement to be RoHS/WEEE compliant merely meant that all the banned substances needed to be excluded from the product and the burden would fall on their suppliers to leave the banned substances out. The reality of lead-free assembly is much more difficult to meet. In assembly, the boards must reflowed to much higher temperatures for longer times. Therefore, the material and curing strategies used in PCB cores and prepregs are rendered obsolete, and alternate strategies are required for meeting flame retardant requirements, discussed in the next section.

8.2.1 Materials for Higher-Temperature Use, Including Halogen-free

Dicyanamide is a latent curing agent for epoxies commonly used before the advent of lead-free capable materials. High glass transition temperature (T_g) materials made with this curing agent did not have adequate temperature capability for lead-free assembly on high-layer-count boards, although it was adequate for some low-layer-count applications. Multifunctional epoxies with phenolic curing agents were adopted to provide higher-temperature capabilities chemically. In addition, low coefficient of thermal expansion (CTE) inert fillers were compounded with the resins to reduce the overall

Active or Pending Legislation in the USA & Canada

Federal

FED		
	HR 4076	Pending
	S 510	Pending
	HR 425	Pending
	HR 666	Pending

State

State	Bill	Status
CA	SB20/SB50	Yes
	AB2901	Yes
	AB302/2587/263	Yes
	AB1415	Pending
	AB1125	Pending
	SB1180	Pending
CO	HB 1256	Pending
CT	HB 6426	Pending
	SB 785/793	Pending
	Hg Edu Red Act	Pending
FL	SB 674	Pending
	H1065/SB1534	Yes
HA	HR 2013	Yes
	HR 170	Pending
	SB 29/1077	Pending
IL	HB 1165/1149	Pending
	HB2572/SB424	Pending
	HB2346/SB1720	Pending
	SB 773	Pending
	SB 2551	Yes
LA	SCRG & 853	Pending

State	Bill	Status
MA	HB 1533	Pending
	S692/H2482	Pending
	HB 1949	Pending
MD	HB 109	Yes
	HB 575	Yes
	HB 83	Yes
	HB 136	Yes
ME	FL 661	Yes
	LD 143	Yes
	LD 1790	Yes
	PL 2001	Yes
MI	SB 583/218	Pending
	HB 429/4150	Pending
	SB 147/158	Pending
	HB 4406 & SB14	Yes
	HB 4036/405	Pending
MN	F 909 (Sec 12)	Yes
	HBA (SB4A)	Pending
	HB 1299	Pending
MT	S/R 15	Pending
NC	H1878 & S 970	Pending
	SB 1030	Pending
NE	NB 301	Pending
NH	HB 73	Pending
	HB 562	Pending
NM	HJM	Pending
	SJM 9	Pending

State	Bill	Status
NV	AB 65	Pending
NY	S890 & A3633	Pending
	A9302	Pending
	A6096	Pending
	A 10050-A7621	Yes
	Chp145 title27	Yes
NYC	NYC BW	Pending
OH	SB 274/SB 70	Pending
OR	SB 867	Pending
	HB 2971	Pending
	HB 3563/SB 740	Pending
	SB 962	Pending
RI	HR 9829	Pending
	SB 234/178	Pending
	HB5783/SB826	Pending
	Info	Yes
SC	SJR 148	Pending
TX	SB 1238/HB 2387	Pending
VA	HB 2378	Yes
	HB 1317	Pending
VT	HB 343	Pending
	S 0111/SB 84	Pending
	TL10 Cp159 6621	Yes
WA	HB 2488	Yes
	HB 1942	Pending
	Exec. Order 04-01	Yes
	HB 1002	Yes

Canada

AB	AR 94/2004	Yes
NS	Sec 102 EnvAc	Pending
ONT	RA04E0018	Pending

FIGURE 8.1 Summary of active or pending legislation in the United States and Canada. [Courtesy of Dynamic Details Incorporated (DDi), Anaheim, Calif.]

275

change in dimension during thermal excursions. This provided higher-temperature capabilities mechanically.

Flame retardant materials interfere with the burning process, for example, either by cooling or by preferentially reacting with decomposition products, excluding oxygen. As additives, their intent is to reduce the risk of injury from fire. Printed-circuit boards were formulated using brominated hydrocarbons and antimony trioxide to ensure they burned slowly and were self-extinguishing. Certain of the sources of bromine, which includes most polybrominated diphenyl ethers and polybrominated biphenyls, have been found to have unacceptable environmental risks: they are bioaccumulative and have been banned by RoHS and other environmental laws. The chief source of bromine in current PCB materials is tetrabromobisphenol–A (TBBA).

TBBA reacts into the backbone of the epoxy polymer, but limits the temperature capability of laminate materials where it is incorporated. It is intended to evaporate and decompose at temperatures that may allow flammable gaseous decomposition products to burn. Its presence interferes with combustion. This occurs above the melting point (180°C), and it is generally accepted that it occurs at a relevant rate above 230°C if unreacted. Since lead-free assembly reflow processes can peak at 260°C and higher, the quantity incorporated into the formulation needs to be balanced with the functionality of the epoxy used.

TBBA seems to be guilty by association with "bad" bromine containing flame retardant additives rather than by a body of evidence. The perception that bromine is bad has led to elimination of any source of bromine in halogen-free materials.

Halogen-free laminates have less than 900 ppm combined chlorine and bromine from any source. For the purpose of this definition, there are no restrictions on other halogens: fluorine, iodine, or astatine. Fluorinated hydrocarbons, such as polytetrafluoroethylene (PTFE), are used in high-frequency applications but could be considered halogen-free in this definition. Iodine is rarely used as a flame retardant, and astatine rarely occurs in nature.

Removing the bromine required a change in strategy of formulating core and prepreg materials to ensure flame retardancy. A common halogen-free strategy is to incorporate materials that foam the board when it is exposed to heat and begins to melt. The foam creates a glass or char that insulates and acts as a gas barrier. Phosphate esters and melamine polyphosphates are representative of proprietary flame retardants compounded into the halogen-free laminates. These flame retardant additives affect thermal, electrical, and physical properties of the cores and prepregs produced.

NanYa NPG-170TL, Hitachi BE 67G, Taiwan Union TU-642, Nelco N4000-7EF, and Isola IS500 represent the industry's efforts to date to find a combination of acceptable properties.

8.2.2   RoHS-Compliant Board Finishes

The most common finish, hot air solder leveling (HASL), using eutectic tin/lead solder is banned. The question is, Which is the best replacement in a lead-free assembly environment?

- Lead-free HASL
- Immersion tin
- Immersion silver
- Electroless nickel immersion gold (ENIG)
- Direct immersion gold (DIG), electrolytic gold
- Organic solder preservative (OSP)
- Other specialty finishes

Each of the finishes has its problems, strengths, opponents, and proponents as discussed below:

HASL is applied to the boards with hot solder and then high-velocity hot air impinging on the surface to remove excess molten solder. The process leaves a film that protects the board circuits' copper surface for later assembly operations. The thickness of the HASL from pad to pad and within the pad varies. Tin/lead is the most common form of solder used in HASL and is not green. This finish is well known in terms of reliability and has often been specified in applications where the product life cycle is as long as 20 years such as for computer servers, has life support requirements, or is a military application. Therefore, this finish has proponents in the high-reliability segment of the market. The electronics industry has been trying to move away from this finish to minimize lead's finding its way into the environment when the boards are scrapped.

Lead-free HASL suffers from more thermal exposure than the traditional tin/lead HASL. Thickness is less variable, and it is more smooth and uniform. It is generally not as visually pleasing (shiny), with unproven reliability in many instances. Currently, not many North American fabricators have adopted this technology. Tin alloyed with copper, cobalt, nickel, and germanium is commercially available and have been found to be compatible with the component lead-free finishes. This finish will be a good candidate for lead-free assembly of lower-layer-count boards.

Immersion tin is highly solderable but may have limited shelf life in warm, humid environments. Thickness control and coplanarity are very good. Pure tin is known to be susceptible to a growth called *tin whiskers* that are produced when the finish is stressed. Tin whiskers can bridge short pads and traces. Many board fabricators advise their customers not to use this finish due to the shelf life and potential reliability issues. In practice, most of the tin mingles with the solder during assembly. This contaminated tin is less susceptible to whiskers than pure tin.

Immersion silver is highly solderable, has good shelf life of up to 12 months, has good thickness control and planarity, and is very compatible with the SAC lead-free solder alloys. It is currently the preferred finish for many OEMs. It is reworkable and may be wire-bonded. Process control is important to prevent voiding at the copper-solder interface during assembly processes. Variables that need to be controlled include copper surface roughness, process chemical composition, concentrations, and temperature.

ENIG is a two-layer finish. The first layer of nickel prevents copper from moving into the gold. The second thin layer of gold prevents the nickel from oxidizing. During the solder operation, the gold dissolves into the solder, and solder wets the underlying nickel. Rework is more difficult because the nickel is brittle. If the chemistry is off, or if the nickel is allowed to form an oxide, the pad either will not solder or will have a defect known as *black pad*. Nickel is not as conductive as the other metallic finishes and has magnetic properties that may be undesirable in some high-speed electronic circuit applications.

Direct immersion gold and electrolytic gold are also appropriate for lead-free assembly. DIG may be cost-effective even compared with OSP and has most of the advantages of the other immersion processes. Electrolytic gold is sometimes used as an etch mask and is common for wire bonding and connectors. Electrolytic gold is generally not an economic choice, unless another function in addition to solder attachment is required.

OSP is a good choice for single-sided applications, but may be a problem with boards containing both SMT and through-hole components and requiring two-sided SMT reflow and a wave solder. Specific high-temperature versions have been developed. Some OSP and flux combinations work better than others when using OSP board finishes.

8.3   Impact of Electronic Design

Modern printed-circuit boards are designed to provide electrical interconnects between active and passive devices in the minimum amount of space. Control of the board's electrical properties is important because it affects the function of the electronic circuits. The board is an electrical component with resistance, capacitance, and sometimes inductance. Unfortunately, the green materials do not always have exactly the same electrical characteristics as the materials they are replacing. Some differences can be accommodated with altering line width and spacing, changing the distance between conductive planes, and modifying the percentage of resin to glass in the cores and prepregs. Unfortunately, design actions optimizing electrical properties may not result in the best performance of thermal or mechanical properties. In a typical multilayer printed-wiring board, stresses due to differential expansion during processing in a lead-free assembly are amplified by changes in thickness and layer design. Figure 8.2 is a detailed typical printed-wiring board (PWB) construction.

8.3.1 Green Material Selection and Test

It is always desirable to define the requirements of a laminate with reproducible laboratory tests that can be performed on different test instruments with similar and valid outcomes. For lead-free assembly of green laminates, the electronics industry has identified several test methods that, when taken together, can evaluate the characteristics and robustness of laminates. These include the following:

- The z axis coefficient of thermal expansion (CTE) (lower is better)
- Glass transition temperature T_g (higher is better)
- Thermal decomposition temperature T_d (higher is better)
- Time to delamination at 260, 288, and 300°C (T260, T288, and T300) (longer is better)
- Toughness of the composite as a function of temperature
- Resistance to cathodic/anodic filamentation (CAF) growth
- Interconnect stress testing (IST)
- Highly accelerated thermal shock (HATS)

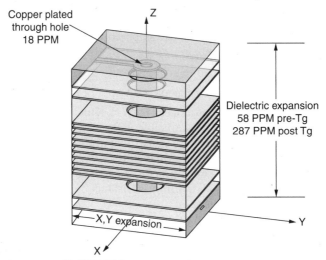

- Stress increases with PWB thickness
- Stress increases as process temperatures due to differential expansion

FIGURE 8.2 Typical printed-wiring board construction. (Courtesy of DDi.)

8.3.2   Green Fabrication Processes

For the most part, green boards are processed following similar flows to conventional laminates. Greater attention must be paid for certain steps to prevent latent failures in subsequent processes or in assembly. Sequential lamination for a three-dimensional multilayer board containing buried and blind vias illustrates some of the issues that have to be addressed, as shown in Fig. 8.3.

Compatibility with Green Materials

Green materials have necessitated a review of the entire board fabrication process. The following is a summary of the processes impacted by green materials and steps taken by the board fabrication industry to remedy potential problems:

1. *Sequential lamination process.* In this process, the materials of construction are subjected to multiple heat exposures and sometimes unequal thermal cycles. In the case shown in Fig. 8.3, one sublamination layer is subjected to four times the heat cycles of its adjacent layers in a lamination press. Most materials shrink as they cure, resulting in additional stress on the surrounding materials and previously cured prepreg. While laminated boards might be considered as cured, additional chemical reaction takes place with each thermal excursion. Depending on the raw materials, several strategies have to be used to ensure that the finished product is reliable and contains minimal latent defects. These include the following:

   a. Optimized lamination cycles for each sequence (temperature ramps, time at temperature, vacuum time, pressure, cooling rate, etc.)

   b. Heat soaks to remove moisture prior to lamination

   c. Stress-relieving steps in and out of the lamination press

   Each fabricator needs to work with the laminate suppliers to ensure that a compatible material and process is used for green laminate board construction. If the lamination process is not optimized, residual stresses in the board are exaggerated in the assembly process, and the board may stress-relieve itself as delamination, high-resistance cracks, and other defects resulting in board failures.

2. *Hole generation.* Drill life on the FR4 materials capable for lead-free assembly has been reduced compared to that for the more conventional materials. The major drill suppliers are working with board fabricators to identify drill bits, drilling parameters, and procedures which provide cleaner holes for subsequent processing. Many of the common laminate materials are filled with ceramic particles to minimize z axis growth during thermal excursions. These fillers affect how

Sequential lamination
stacked Via

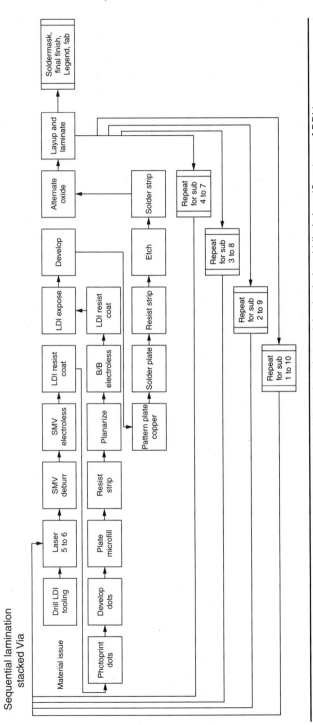

FIGURE 8.3 Sequential lamination for three-dimensional multilayer board containing buried and blind vias. (Courtesy of DDi.)

the material chips and cracks during drilling. When laser-generated vias are drilled, tool files may have to be modified to obtain a similar hole if all other factors are the same (such as glass, percentage of resin, and thickness). Most of the defects created at the drilling operation can be visually observed with a microscope, and therefore optimizing the drill and laser process can be performed quickly. Subsequent operations after drilling such as permanganate and plasma desmear appear to behave similarly with the higher-performance green materials as compared to the legacy standard nongreen materials.

3. *Alternate oxide.* A third area of concern concerns the choice of alternate oxide on etched inner layers. Some of the older proprietary processes leave an oxide with less bond capability. Shops using these chemistries will need to change to a more modern oxide to ensure good inner layer adhesion to prepreg bond.

4. *Final surface finish.* Currently, some high-reliability applications are exempt from the ban of lead in solder, and therefore tin/lead HASL can be used as the final finish. Otherwise, OSP, immersion silver, immersion tin, DIG, and ENIG finishes discussed earlier in this chapter are preferred. The residual moisture in these boards needs to be controlled to minimize stress on the board during assembly. The finishes are wet-processed but may deteriorate with long drying cycles. Each fabricator needs to determine how the finished boards will be dried prior to packaging and shipping to customers.

Process Capability Needed for Green Materials

The lead-free assembly requirements have driven the PWB polymer systems to be more resistant to temperature excursions. Formulators have become more inventive in reducing CTE with additives and better curing systems. New flame retardants have been incorporated with regard to synergy with the resin systems. These materials are much more capable than their predecessors if they were assembled with leaded solder.

Process capability is based on not only the inherent performance of the process, but also meeting industry performance requirements. The lead-free assembly process is more severe than the legacy systems and imparts greater stress on the board. The use of careful processing in fabrication steps to enhance the performance of board materials is required, since the performance of some board materials is lacking. Poorly controlled processes include lack of proper tolerance on process parameters, rework on intermediate steps, inclusion of bakes, dull drills, and poor handling, which can have an adverse effect on the quality of the boards.

Required Changes for Successful Green Material Processing

Below is a list of required changes to ensure success in green material processing:

1. Confirm lamination process for each sequence.
2. Incorporate bakes to control moisture during processing.
3. Confirm alternate oxide compatibility.
4. Confirm drill and laser parameters.
5. Tighten process controls throughout.

8.4 Material Screening

8.4.1 Processability of Green Materials

Candidate green materials need to behave in a similar methodology as current materials to be compatible with the legacy equipment set and processes such as chemical resistance, drill parameters, lamination cycle, and desmearing.

8.4.2 Compatibility with Assembly Processes

Lead-free assembly requires tighter controls on the assembly reflow process because thermal constraints reduce the size of the process window. In Fig. 8.4, two curves represent the solder reflow in a typical SMT oven. The bottom one, tin/lead, has a sharp peak after a soak at lower

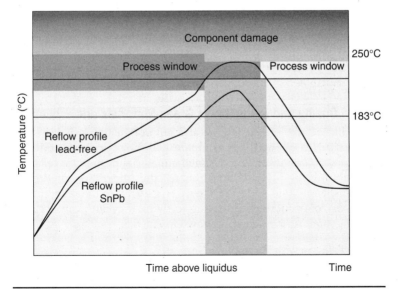

FIGURE 8.4 Solder reflow profile in a typical SMT oven. (Courtesy of DDi.)

temperature. The top profile (lead-free) also has a soak, at higher temperature, followed by equilibrium and longer hold time. The process is bounded at the top at 250°C by the capability of components to survive the process. This is a fixed temperature (hard stop) value due to physics at the silicon where the functionality of active chip components is changed and is not reversible. Under this hard stop, there is a process window block between the peak temperature of the tin/lead reflow peak and the hard stop, shown to the left of the peak. To the right there is a similar, but significantly thinner, block between the lead-free reflow peak and the hard stop. This thinner process window requires that the lead-free assembly have tighter controls on the process.

During SMT reflow, the board is conveyed through an oven with several temperature-controlled zones designed to rapidly bring the board up to temperature while not activating the flux prematurely. The ramp rate is limited by the flux and is bounded by both a shallow and a steep rate. The packages and discrete devices to be attached shadow the board temperature according to their cross-sectional area and different mass variation from component to component. Nonuniform mass and shadowing tends to cause the board to have a differential, asymmetrical temperature across it. To make sure all the components reach the reflow temperature of the solder, some parts of the board have to be much hotter. The differential worst case can be as high as 30°C. This results in stress in the board due to CTE movement point to point on the board. To help manage the differential, ovens have more zones, are longer, and may have to run slower. The increased temperature and dwell time challenges the performance of the board. More layers and thus thicker boards are more highly stressed as the temperature differentials in the board are more severe.

A thermal excursion in the assembly process can cause defects that normally would be attributed to poor board thermal performance. The challenge in finding compatible lead-free PWB materials lies in finding alternatives with robustness in the temperature "excursions" anticipated in subsequent manufacturing.

8.4.3   Quality and Reliability of Green Materials and Processes

A number of accelerated fatigue and environmental tests are commonly used to qualitatively compare materials, designs, and processes and to quantitatively describe a material property over time. For a multilayer printed-circuit board, the integration of design, materials, and value-added processes is reduced to the reliability of the z axis connections. Failure due to thermal cycling can occur in several geometric locations, as shown in Fig. 8.5.

Typical Failures with Thermal Cycling

When a circuit board is powered and then turned off and allowed to cool, the circuit board materials—glass, resin, copper, and plating finishes—do not expand and contract at the same rate. The resin

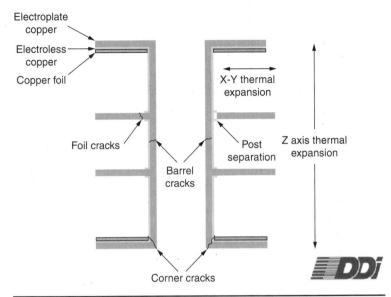

Electroplate copper

Electroless copper

Copper foil

X-Y thermal expansion

Z axis thermal expansion

Foil cracks

Post separation

Barrel cracks

Corner cracks

FIGURE 8.5   Typical failures due to thermal cycling. (Courtesy of DDi.)

component in the board expands at a much higher rate than the copper in the plated through hole. The copper resists the movement of the resin in the laminate and is stretched slightly during each temperature excursion. Pure copper has very little elasticity and deforms on stress. With this deformity, the stress may be relieved mechanically with cracks at weak points. A typical plated through-hole construction is shown in Fig. 8.5. During a thermal stress, caused by either turning the circuit power on or off, or sequential solder during assembly, the board will grow and shrink in each direction. Failures in the conductive portions can result in lost or diminished function of the board. These failures can be in any of these forms:

- Corner cracks
- Barrel cracks
- Foil cracks
- Post separation

More descriptive names for failures are often required to help distinguish their causes, but these are typically subsets of the names above. Figure 8.6 is a photomicrograph of a failed board which illustrates the failure issue.

Development of materials for the more severe lead-free solder conditions requires a balance of material properties that challenged the formulators:

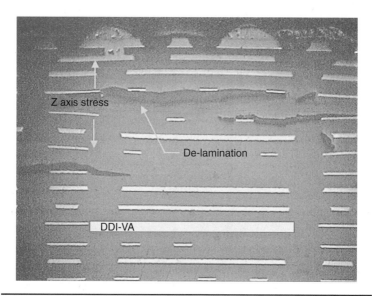

Figure 8.6   Photomicrograph of a failed board. (Courtesy of DDi.)

1. The dimensional changes above and below the glass transition temperature T_g of the multilayer laminates needed to be significantly smaller than in the historical laminate materials to reduce stress on the plated through hole during the higher-temperature solder processes.

2. Adhesion to plated through-hole barrels, foils, and reinforcements had to be greater at extreme temperatures to resist delamination.

3. High, void-free adhesion of prepreg to etched core, conductors, and spaces had to be increased.

4. Resin systems needed to have higher thermal decomposition temperature T_d to prevent local void formation adjacent to the plated through hole.

5. Resin systems needed to be able to withstand higher temperatures longer.

6. Resin systems needed to be robust in the presence of moisture, either by not absorbing moisture or by not being affected by it, in terms of having better moisture absorption properties.

7. New materials needed to continue to meet flame retardant requirements to UL 94V0.

8. New materials not cause significant changes in the material and processing costs of the raw laminate and prepreg.

Preconditioning

As a typical board fabricator, DDi of Anaheim, California, simulates how its products are to be used in downstream processes during evaluations, as in Fig. 8.7, showing a typical lead-free reflow profile. This profile can be used on many lead-free multilayer PCBs that incorporate a combination of BGA, discrete SMT, and through-hole components on both sides of the board. This profile changes with board design, density of components, and other factors known to the PCB assembly industry. Customer product use might require certain conditions that DDi will incorporate into their test plans when evaluating new materials or processes, prior to reliability testing. As indicated in the following sections (IST and HATS), this simulated solder reflow is included as prior history before the PWBs undergo reliability testing in the form of thermal cycling. Note that preconditioning for both (HATS and IST) sections are prescribed. Conditioning in an actual assembly process is not normally required for these tests, but has been included in some of the company's evaluations.

The thermal cycle shown in Fig. 8.7 represents a mild reflow condition for lead-free applications. This represents an assembly operation that is focused on a robust process in providing a reliable product. This profile may not be as fast as some assemblers would like to run

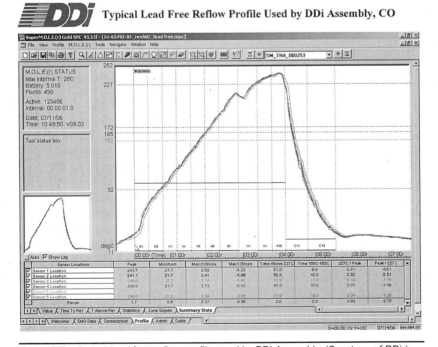

FIGURE 8.7 Typical lead-free reflow profile used by DDi Assembly. (Courtesy of DDi.)

their lines, while others might have a wide variation in the severity of the reflow profiles. These variations include the following:

1. Slope of temperature ramp that is consistent with the limits associated with the flux used in these systems
2. Time above liquidus (TAL), which is 217°C for solder alloy SAC305, in ranges of 30, 60, and 90 s
3. Peak temperatures reached during lead-free solder wave and reflow (245 and 260°C)

The combination of these profiles represents the operating envelope of many boards during assembly processes.

Multilayer boards typically go through reflow three times during a normal assembly operation: one SMT pass for each side and once through a wave for through-hole components. In some cases, a BGA component may be removed and replaced (reworked), once or twice. In this case, the board sees an additional two reflows of TAL cycles for each rework. Worst case, a board could see seven reflows. Each reflow degrades the FR4 laminate to some extent. Degradation may be expressed as undesirable recombination, cross-linking, bond cleavage, or oxidation that compromises the intent of the material design, which may also reduce the usefulness of the board for its intended use. For more critical applications, this effect can be assessed with reliability testing. In the electronics laminate industry, several tests for reliability are used, including the acronyms IST, HATS, and CAF explained below.

Interconnect Stress Test (IST)

The IST was developed in the 1990s to overcome problems with air-to-air temperature cycling in testing PWBs. The test vehicles are specific coupons taken from the PWB with plated through holes. In dual chamber, air-to-air cycling, the coupons are moved from one environment to another. If the coupons are to be monitored during test, umbilical power and signal cords also move with the coupons. A 1000-cycle test could take up to 42 days and can give an indication of the reliability of the electrical interconnect and harness, as well as the device under test.

As part of the IST, an instrument provides power to internally heat the plated through holes (PTHs) by DC resistance heating and uses a second sensing circuit to monitor the resistance of the PTH. The coupons are cycled between ambient and an elevated temperature. Heating from within simulates turning a power device on and off and the heat flow from the device into the board. There is an option to incorporate features in the coupon that will allow the capacitance of the circuit to be monitored as well. Capacitance is more sensitive to changes in thickness due to expansion and delamination that may occur than the corresponding change to the resistance of the plated through holes.

The field performance of electronic circuits could be simulated by testing at several temperatures to cycle coupons to failure. Higher cycle temperatures simulate longer field use times. IST can be cycled to higher temperatures than the HATS example discussed below. HATS tests can also quench the test coupons to –40°C. There are many companies that specialize in these reliability tests, including PWB Interconnect Solutions, Inc., of Ottawa, Ontario, Canada. Many papers were published by them and others that present calculations of life acceleration factors using IST equipment. In practice, most board manufacturers and their customers consider that if coupons pass 300 IST cycles, then they are of sufficient quality for normal product use. Preconditioning inside the chamber, or prior to entry into the chamber, and the maximum cycle temperature will have a measured effect on the performance of the coupons.

Preconditioning in the IST oven to simulate lead-free assembly processing offers an opportunity to predict how the board will act after components have been assembled and the board is being actively used. An example of test documentation for an IST is given in Fig. 8.8.

Highly Accelerated Thermal Shock (HATS) Test

The HATS test is similar to dual-chamber thermal shock techniques but is much faster and easier for monitoring samples during test. In the HATS chamber, specially designed coupons are placed in an airflow that is sourced hot or cold to shock the specimens. The specimens are stationary with 4 daisy chain nets per coupon. This allows 144 nets to be tested at once compared to 12 in IST. The mass of the coupon is brought to temperature with less temperature differential within the board than occurs in a similar IST.

Companies such as DDi use both IST and HATS coupons to evaluate materials, process changes, and design features that may affect reliability of boards. OEMs often also use these techniques to evaluate their board suppliers. Both techniques have been used to predict reliability and produce similar failure modes as those encountered in the field. In some cases, historical reliability recording requires continued use of tests such as IST or HATS to provide baseline comparisons against future material and process changes.

Temperature range								
Low		High		Precondition conditions			Number of specimen	Stop test
Ambient	150°C	170°C	190°C	230°C	3x	6x	24	1000 cycles
				260°C	3x	6x	24	

FIGURE 8.8   Typical IST documentation.

Both HATS and IST stress the z axis interconnects by differential expansion and contraction of the conductive versus the nonmetallic media surrounding vias or plated through holes. For predictive analysis to be made from the data, the failure modes of the accelerated tests should mimic field failures in both type and frequency. To achieve the same temperature range, IST must be cycled to a higher temperature than HATS due to the difference in low-temperature limits of the tests. Higher temperatures tend to create a higher frequency of some failure modes than might be expected in the field. HATS should be used when comparing material properties, where significantly more coupons under test should be used with the same conditions of thermal cycling and lower peak temperature. An example of test documentation for a HATS test is given in Fig. 8.9.

Cathodic Anodic Filamentation (CAF) Test

The CAF test produces growth that is a subsurface failure expressed as a high-resistance short following glass bundles in the laminate. A CAF test is conducted at high temperature and humidity over a period of up to 1000 h with a DC voltage between adjacent structures common in printed-circuit boards. The tests are conducted on specially designed coupons that control distances between structures being investigated.

CAF formation is a deposit of insoluble copper salts between adjacent conductive geometries such as vias, traces, planes, and antipads. These salt bridges are not highly conductive and are easily burned out in test; that is, failures appear to heal themselves at high voltage, which could be very disruptive in actual product use (see Sec. 3.31 for more discussions about CAF).

10 Layer board									
HATS			Number of specimens	Low temperature centigrade	High Temperature centigrade	Via size	Stop test*		
Sample set	Process	Precondition							
1	Product under test		12	−40	130	3	>90% Fail		
2	Product under test		12	−40	145	3	>90% Fail		
3	Product under test		12	−40	160	3	>90% Fail		
4	Product under test		12	−40	130	4	>90% Fail		
5	Product under test		12	−40	145	4	>90% Fail		
6	Product under test		12	−40	160	4	>90% Fail		
7	Product under test		12	−40	130	5	>90% Fail		
8	Product under test		12	−40	145	5	>90% Fail		
9	Product under test		12	−40	160	5	>90% Fail		
10	Product under test		12	−40	130	6	>90% Fail		
11	Product under test		12	−40	145	6	>90% Fail		
12	Product under test		12	−40	160	6	>90% Fail		
13	Sequential baseline		36	−40	160	5	>90% Fail		
							*Continue until all sample sets meet criteria		

FIGURE 8.9 Typical HATS test documentation.

Typical test voltages are 10, 48, and 100 V. It is common to run the test at twice the actual use voltage. The growth is thought to be diffusion-limited and therefore has a nonlinear rate of formation. Coupons are monitored at intervals between 96 and 1000 h for resistance.

CAF reliability is the sum of design, materials, and processes for producing the device under test. A number of other effects (e.g., flux residue and other surface insulation resistance factors) can confound the results. In addition, the choice of bias voltage affects the result: too high a voltage will burn out a filament if formed during the test; too low a voltage may not create enough failures to allow comparison.

Because of the interferences to this test, CAF is more difficult to conduct and interpret than IST and HATS tests. There might be unexplained failures in CAF in most test programs that are unrelated to the materials used in the test. To rule out other factors, materials comparisons need to be from boards produced from multiple lots of materials at different time periods. If different PWB suppliers can also be included, reasonable comparisons can be made between the robustness of one laminate material versus another.

Multiple exposures to lead-free assembly temperature profiles challenge the interface between the glass fibers and the phenolic cured epoxy fillers. Some of the early contenders for lead-free assembly capable laminates, especially those formulated for high-frequency applications, did not do well in this test. Some materials failed after 100 h at 48 V, hole wall to hole wall, dropping 5 orders of magnitude in resistance.

Laminate suppliers were forced to go beyond historical formulation techniques to overcome CAF deficiencies. One approach that offered some hope was a discovery that a thin coating on the glass sizing could offer an opportunity to inhibit CAF growth while still maintaining its function at the weaver to lubricate and protect the glass. Proper selection of the sizing helped the strands to be wet in the saturation process and to prevent debonding during the stress of lead-free assembly and subsequent high-temperature humidity aging.

In practice, modifications of the formulations described above in conjunction with compatible sizing on the glass yielded products that did not fail after 600 h at 48 V, hole wall to hole wall. An example of test documentation for a CAF test is given in Fig. 8.10.

Typical Green Material Test Plan

As a typical board supplier, DDi evaluates candidate materials with two representative boards:

- A 24-layer single-lamination server design with IST, HATS, and CAF structures with conventional plated through holes

- A 10-layer sequential laminated phone board with thin cores and prepregs, laser-generated vias, and plated through holes.

Coupons 1–2, 9–10 on 10 layer board								
CAF, modified alcatel coupon								
Sample set		Via size (mils)	Process	Number of samples	Temperature	Relative humidity	Hours conditioning	Test hours	Bias
1	Hole to hole	3	Product under test	25	65C	85	96	500	10
2	Hole to hole	3	Product under test	25	65C	85	96	500	48
3	Hole to plane	3	Product under test	25	65C	85	96	500	10
4	Hole to plane	3	Product under test	25	65C	85	96	500	48
5	Hole to trace	3	Product under test	25	65C	85	96	500	10
6	Hole to trace	3	Product under test	25	65C	85	96	500	48
7	Hole to hole	4	Product under test	25	65C	85	96	500	10
8	Hole to hole	4	Product under test	25	65C	85	96	500	48
9	Hole to plane	4	Product under test	25	65C	85	96	500	10
10	Hole to plane	4	Product under test	25	65C	85	96	500	48
11	Hole to trace	4	Product under test	25	65C	85	96	500	10
12	Hole to trace	4	Product under test	25	65C	85	96	500	48
13	Hole to hole	5	Product under test	25	65C	85	96	500	10
14	Hole to hole	5	Product under test	25	65C	85	96	500	48
15	Hole to plane	5	Product under test	25	65C	85	96	500	10
16	Hole to plane	5	Product under test	25	65C	85	96	500	48
17	Hole to trace	5	Product under test	25	65C	85	96	500	10
18	Hole to trace	5	Product under test	25	65C	85	96	500	48
18	Hole to hole	5	Sequential baseline	25	65C	85	96	500	10
18	Hole to hole	5	Sequential baseline	25	65C	85	96	500	48
18	Hole to plane	5	Sequential baseline	25	65C	85	96	500	10
18	Hole to plane	5	Sequential baseline	25	65C	85	96	500	48
18	Hole to trace	5	Sequential baseline	25	65C	85	96	500	10
18	Hole to trace	5	Sequential baseline	25	65C	85	96	500	48
Environment per IPC TM650 2.6.25								
Analysis								
Weibel each sample set with failures								
Identify any SIR failures								
Section, SEM representative failures to determine if CAF, whisker formation, or other failure mode								

Figure 8.10   Typical CAF test documentation.

A typical evaluation of a new process would include the tests mentioned above: IST, HATS, and CAF. Examples of test documentation are given in Figs. 8.8, 8.9, and 8.10.

8.5 Due Diligence

Due diligence refers to a voluntary effort by companies to validate conformance of their products to prevailing regulations in countries where they or their customers sell or deliver products. Proof of compliance could vary from "trust me" to destructive testing. Due diligence represents a low standard of proof where the premise that the company meets the requirements is more likely than not. With time and additional evidence of compliance, it is assumed that the level will improve. Most companies' approach has been to find a "reasonable" balance of evidence that fits into their existing systems, perhaps providing greater assurance than the industry standard practice without giving up a competitive advantage. An example of such due diligence is the compliance strategy chain of evidence as practiced by DDi in Fig. 8.11.

8.5.1 Design Control for Leaded versus Lead-Free Boards

Many companies will produce boards for RoHS-exempt (leaded) as well as lead-free boards. These exemptions are controlled through the customer drawings, data, and specifications. The presence of these nongreen designs requires controls to prevent escapes. Control can be successfully achieved with unique part numbers.

During the transition to lead-free, HASL board designs which included hot air solder (tin/lead) leveling prior to the EU RoHS directive can be flagged for the customer to revise on reorder. It is common for a single specification to be revised, without causing all other board design documentation to be updated, to achieve a compliant board finish and material set.

These implied additional requirements can be imposed at several levels in the organization:

- Sales
- Preliminary engineering review of design for manufacturability (DFM)
- Specification review
- Quality, R&D, process engineering, and generally anyone having contact with customers

Conditions triggering that boards need to be compliant (capable of lead-free assembly) are discussed in Chap. 6 and include the following:

1. Customers requesting RoHS certificate of compliance and/or material declaration
2. Customers including a note or clause requiring boards or coupons to be soldered with multiple exposures equal to or greater than 260°C, or one exposure exceeding 288°C.
3. Customers explicitly stating that their product will be assembled in a lead-free environment

Lead-Free Assembly Impact on PWB Fabrication

Customer requirements come to board fabricators in all the usual formal and informal manners; by drawings, by inside data and specifications, and in the boilerplate on purchase orders. Fabricators build to print, but are concerned when the material set or design is

294

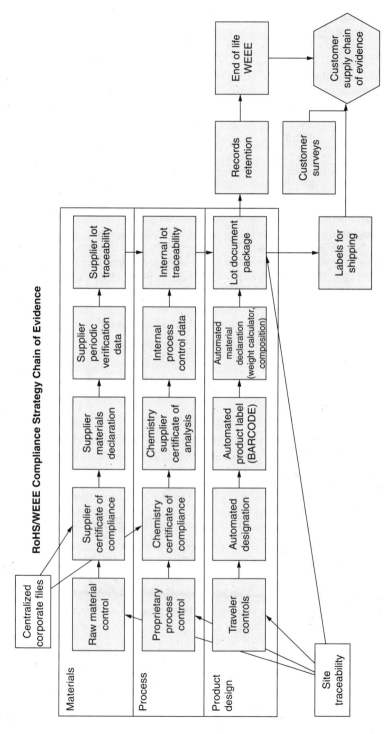

RoHS/WEEE Compliance Strategy Chain of Evidence

FIGURE 8.11 Due diligence example. (Courtesy of DDi.)

not consistent with the assembly process. They need to make sure that the boards will survive after leaving their factories. Unfortunately, current lead-free solder formulations that meet the RoHS directives require higher temperatures to ensure good joints than the equivalent legacy tin/lead solder. The existing FR4 material set did not behave well in all designs. So the fabricators had to conduct extensive testing to evaluate the impact of these changes.

An example would be DDi. They divided lead-free assembly applications into three groups, A, B, and C, depending on reflow profile and board thickness, as shown in Fig. 8.12. The tests included thermal stability, peel strength, solder float, electrical conductivity, and the IST, HATS, and CAF tests. In addition, they evaluated 10 different FR4 materials and made several classifications of FR4 boards by material properties, as shown in Fig. 8.13. These properties include

1. Resin system

2. T_g (glass transition temperature inflection point) where the CTE changes due to increased molecular mobility

3. D_k (dielectric constant), which is the relative ability of the insulator to hold a capacitive charge as compared to free space

4. df (dissipation factor), which is the decimal ratio of the irrecoverable to the recoverable electrical energy introduced into the insulating material, when exposed to electromagnetic field

Other options to be considered when ordering boards include the following:

1. Surface finish type

2. Solder mask type, supplier, and color

3. Legend, including supplier and color

Three major groups for Pb-Free assembly:

Group A: PCB thickness up to 0.062 in
Peak temperature no greater than 245°C
Time above reflow 30 to 60 s

Group B: PCB thickness up to 0.125 in
Peak temperature no greater than 260°C
Time above reflow 60 to 90 s

Group C: PCB thickness up to 0.187 in or greater
Peak temperature no greater than 288°C
Time above reflow 90 to 120 s

FIGURE 8.12   Lead-free assembly classifications for test purposes. (Courtesy of DDi.)

Standard FR-4 group A

Material	Resin system	T_g	D_k	df
Nelco N4000–11	Multifunctional	175°C	4.1	0.018@1 GHz
Isola FR-370 HR	Multifunctional	180°C	4.04	0.021@2 GHz
Nelco N4000-29	Multifunctional	185°C	4.3	0.016@1 GHz

High performance FR-4 group A & B

Material	Resin system	T_g	D_k	df
Nelco N4000–13EP	Enhanced multifunctional	210°C	3.7	0.009@1 GHz
Isola IS620	Modified epoxy	225°C	3.61	0.008@2 GHz

1.

Standard FR-4 group B

Material	Resin system	T_g	D_k	df
Isola FR-370 HR	Multifunctional	180°C	4.04	0.021@2 GHz
Nelco N4000–29	Multifunctional	185°C	4.3	0.016@1 GHz

At this time DDi has not qualified a substrate material capable of meeting the highest-performance group C.

Figure 8.13   Board classifications by material properties. (Courtesy of DDi.) FR-4 = ANSI category for epoxy glass circuit board material including flame-retardant additives.

Within the industry, the IPC has published Standard IPC-4101B with slash sheets to represent generic products which can meet lead-free assembly requirements. Specifically, /126 and /129 represent materials which meet the most rigorous requirements. The specifications address tests and methods, reproducible from one lab to the next, that provide a decision point on the product produced as a raw material for fabricating boards. Board fabricators see this as an *entitlement* (minimum of expected properties), but are more concerned with the integration of these materials, the design, and value-added processes to provide a functional board.

8.5.2   Process Control: Contamination and Moisture Management

Each step of the manufacturing operations has the potential to add a banned substance contaminant to the product. Companies should normally choose to look for potential failures and implement process controls to reduce their risk of occurrence.

Some examples of potential failures include

- Wrong material defined (eliminated by product engineering check)

- Wrong materials released (several preventive controls should be put in place)

- Wrong finish defined in sequence of operation (eliminated by product engineering check)

Less obvious contamination:

- Incomplete strip of solder etch mask: monitor the lead contaminant residual in the holes by extraction and analysis by atomic absorption (AA) spectroscopy or convert to tin etch mask
- Lead contaminations in electroless nickel bath for ENIG finish

As processes are improved, changes are rendering the chemistry and plating processes more difficult, requiring more rigorous qualifications and test data. A good-quality system should be implemented to follow and document all processes of record and manage the traceability for a minimum of 4 years.

Moisture is also important in process control and due diligence: boards assembled in lead-free solder require higher temperatures than similar boards processed with tin/lead processes, resulting in greater influence of moisture in the boards. Boards are subjected to wet processes for cleaning, imaging, plating, bond enhancement, hole preparation, etc.

Fabrication processes after etch are particularly important as the resin-saturated laminate is exposed to process chemistry and may adsorb moisture. Since the next step is encapsulation in a lamination step or solder mask, increased moisture content can interfere with cure, adhesion, or stress in the lamination. In addition, it could create increased stress within the laminate on exposure to soldering processes, resulting in delamination. The vapor pressure of water increases more than two and half times from the tin/lead solder process (205°C) to the maximum lead-free SAC305 process (260°C). Drying steps should be incorporated in the board fabrication process after alternate oxide and prior to solder mask.

Managing moisture during fabrication, particularly before lamination, can leave a board more robust for lead-free assembly operations. Materials should be stored in a controlled environment, and drying and hold times should be monitored.

Providing stress-free boards helps make the assembly process more robust. Stress is produced by lamination, and the lamination cycles for lead-free materials have been similar to past FR4 cycles with a few notable exceptions. With some board designs and materials, the latter part of the lamination cycle is designed to provide stress relief in the board by holding at temperature, at reduced pressure, in a process called annealing. This is particularly true with sequential lamination processes. In sequential lamination, boards are laminated as subboards as many as four or five times in a press. Early laminations have greater stress than later laminations. The annealing process helps reduce the stress contribution of the lamination process.

A conservative assembly process should maintain the boards in a dry environment prior to application of solder paste and placement of components.

8.5.3   Supplier Control

A compliance strategy for meeting green regulations relies heavily on assurances and data supplied by the supply chain. In printed-circuit board manufacture compliance requirements, the suppliers can be divided into three groups: packaging, materials, and chemistries. Annually, as well as on product qualification, the suppliers are solicited for certificates of compliance, disclosure of composition, and similar evidence that banned materials are not included in their products.

The annual solicitation contains the following sections:

- An explanation of how the purchasing company uses the supplier product in its own products
- The pertinent parts of any regulations the company is trying to meet (RoHS, WEEE, and other regulations)
- A request for a certificate of compliance and material declaration
- A list of products needed for the certifications
- A thank you note for the suppliers' help

Packaging has been at the forefront of recycling for many years and is easiest to manage. The fabricator letter to packaging suppliers requires only a certificate of compliance. With the maturing of several offshore regulations (China, for example), the packaging requirements will be more strict. The fabricator might ask for a materials declaration for materials consumed to produce the product, such as laminates, prepregs, adhesives, and solder masks. Chemistries that are used for value-added processes, such as plating and surface finishes, are required to state the expected composition as processed and to state compliance of the banned chemistry substances.

A board fabricator should include a statement on their purchase orders that requires their suppliers to notify them if a product shipped does not meet specific environmental regulations. It may not be adequate to control all possible materials, but this procedure adds another safety net layer to prevent noncompliant materials from entering the process stream. The system for controlling green materials should be similar to all other materials purchasing, and not a stand-alone system.

8.5.4   Records and Material Declarations

Records for each lot processed need to be tracked to materials of construction compliance, process compliance, and design compliance. These records currently need to be maintained for 4 years.

In some markets, the traceability requirements needed for demonstrating green compliance are well prescribed and can be easily modified to include information related to RoHS and other environmental laws. The volume of records has made it preferable to use digital records for primary storage in an accessible database.

Printed-wiring board manufacture requires support of the customer's due diligence to meet environmental laws. While there have been efforts made to standardize reporting, these came late for RoHS. As a result, there are many unique methods of reporting the composition of boards produced both in format and in required data. Until this is sorted out by the industry and their customers, each customer has to be addressed according to his or her needs.

8.6   Green Laminate Economics

Initially, the cost of raw materials complying with the green environmental laws was approximately 10 percent higher than for conventional materials. As the market matures, this cost differential will be nullified.

Value-added operations for green materials are similar in cost with the exception of finishes. Hot air solder leveling (HASL) with tin/lead solder was less expensive than the finishes appropriate for lead-free assembly. A lead-free solder hot air leveling process is gaining acceptance and may allow a lower-cost finish for boards where HASL is appropriate.

Records retention and traceability have been good practices for many applications and markets, but will be expanded to include all products for export from the United States.

8.7   Summary

The electronics industry is continuing a journey incorporating best practices for the environment. Currently, the EU directives and other environmental laws are at the forefront in the evolution of green laminates and processes. As the global society learns more about the effects of our actions on the environment, we can expect the green challenges to escalate as well.

Solving environmental issues requires diligence in ensuring that downstream processes are not compromised. The change in the basic solder interconnect from tin/lead to lead-free is an example of a "solved" environmental issue. Restricting the use of lead in PWB has required significant efforts by material suppliers to reformulate basic products and by fabricators to tighten their processes to allow the new materials to be manufacturable for subsequent assembly. It will take some time for our confidence in real-time reliability of lead-free electronics to reach the level we enjoyed after half a century of electronics experience with tin/lead solder.

Acknowledgments

My employer, DDi, has spent several years developing green PWB at its locations in North America. This chapter represents a consolidation of this work. A number of colleagues in DDi made individual contributions to this chapter that I would like to acknowledge, including Gil White, Tom Buck, Monte Dreyer, and Raj Kumar. They performed many of the supporting test designs and evaluations presented here. Most of the illustrations and tables were a result of their efforts. Wendy Boger, John Chelberg, and Emad Youseff participated in many of the earliest green efforts with industry consortia and organizations. Jon Good and Tony Hall of Veritek Manufacturing Services (formerly DDi) were indispensable, offering their expertise and support in evaluating materials through their SMT assembly operation. Their efforts continue to help the industry meet the tradition of good stewardship in the environment.

References

Bertling, S. "Evaluating Laminates for High Temperature Assembly." *The Board Authority*, pp. 20–23, January 2005.

Brewin, A., L. Zou, and C. Hunt. "Susceptability of Glass Reinforced Epoxy Laminate to Conductive Anodic Filimentation." *NPL Report_MATC(A)155*, National Physical Laboratory, Teddington, Middlesex, United Kingdom, January 2004.

Buck, T., and G. White. "WEEE & RoHS Directive, What Does It Mean?" DDi sales presentation, Anaheim, Calif., 2007. www.ddiglobal.com

"Directive 2002/95/EC of the European Parliament and of the Council of 27 January 2003 on the Restriction of the Use of Certain Hazardous Substances in Electrical and Electonic Equipment." http://europa.eu.int.

Furlong, Jason, and Michael Freda. "Advanced Testing Using Real Life Evaluation and Statistical Data Analysis". White Paper, PWB Interconnect Solutions, Inc., Ottawa, Canada, 2006.

IPC-4101B, "Specification for Base Materials for Rigid and Multilayer Printed Wiring Boards" with amendments, IPC, Bannockburn, Ill., March 2007. www.ipc.org

IPC-1752, "Materials Declaration Management" IPC, Bannockburn, Ill., pp. 1–30, February 2006. www.ipc.org

"Material Composition Declaration for Electronic Products JIG-101." Electronics Industries Alliance (EIA), JEDEC, and the Japanese Green Procurement Survey Standardization Initiative, April 2005. www.eia.org/jig

Neves, R., B. Rick, and A. Timothy. "Highly Accelerated Thermal Shock Reliability Testing." White Paper, Microtek Laboratories, Inc., Anaheim, Calif., 2005.

CHAPTER 9

Green Finishes for IC Components

Donald C. Abbott, Ph.D.

Texas Instruments Incorporated, Dallas, Texas

9.1 Introduction

The need for green finishes, or more specifically lead-free finishes, for peripherally leaded components and array substrates is widely known, being required by RoHS legislation coming out of Europe and China. Redefining the need to be lead-free as opposed to "green" is a consequence of the other five RoHS substances—mercury, cadmium, hexavalent chromium, and the brominated flame retardants PBB and PBDE—never being used in integrated circuit (IC) component finishes. This chapter considers lead frame finishes that become the solderable terminations or leads of IC packages.

9.2 Background

There is a long history of soldering with tin/lead. What was lost in the rush to formulate lead-free finishes was the impact the higher melting-point lead-free solder would have on the printed wiring board (PWB) assembly process and the demands it would make on IC components. The IC manufacturers initially focused on how to remove lead from their product; later they realized the higher reflow temperatures required for PWB assembly with lead-free solders would place greater demand on the moisture sensitivity level (MSL) performance of the IC package.

The transition from non-RoHS compliant brominated flame retardants to RoHS-compliant versions occurred in parallel to the drive to get the lead out. Discussion of the removal of brominated flame retardants for RoHS compliance is beyond the scope of this chapter, as it is fundamentally a mold compound issue that has tangential impact on mold compound adhesion to the lead frame.

Tin/lead soldering has had a long and well-documented history for electronics assembly, going back more than 50 years. During the tin/lead soldering era there was an evolution of IC termination finishes, discussed below.

9.3 Lead Frame Finish Evolution

9.3.1 Gold (Au) Finish

In the early days of IC manufacturing, the late 1960s to middle 1970s, IC leads were finished with relatively thick gold plating, using the wisdom of the day that "nothing solders better than gold." The first gold-plated lead frames were flood (entire surface) plated. The gold also provided an easy surface to make a Au/Au weld for wire bonding. The lead frames were stamped from iron/nickel (FeNi) base metals, such as alloy 42 or Kovar. The coefficient of thermal expansion (CTE) of such alloys closely matches that of the silicon in IC chips. The chips were large, so it was important to minimize CTE mismatch to prevent cracking of the IC. The modest thermal and electrical demands of early IC's were easily met by high iron/nickel base metals. Toward the end of the gold finish period, selective or spot gold, where the gold is plated only in the functional areas of the lead frame , was developed as a cost reduction. Over time, two realizations caused the move away from the gold finish:

1. Gold embrittlement of solder joints was elucidated. Solder joint embrittlement occurs when the gold in the solder joint approaches an upper limit of approximately 3 percent (w/w). Tin/gold (SnAu) intermetallics segregate in the solder and become weak points where the solder joint fractures.

2. Economic considerations became relevant. As ICs became more ubiquitous, the cost of gold flood plating or even spot plating the lead frame with a thick layer of gold became prohibitive.

9.3.2 Silver (Ag) Flood Finish

As a replacement to the full or spot gold finish, a silver flood plated finish was developed for use on Kovar® and alloy 42 base metals. The key to this finish was a strike plated layer prior to the flood plated silver. This strike layer was co-deposited copper (Cu) and silver that progressed from pure copper at the base metal surface, through a decreasing percentage of copper, to pure silver at the surface of the strike. The copper at the base metal surface provided excellent activation and adhesion for subsequent plating. The pure silver at the surface of the strike was an ideal substrate to plate thick silver, since there was no risk of immersion or autocatalytic plating of silver. Despite being a flood plating process, this was inexpensive relative to gold plating and gave excellent wire bond and solderability performance. The Achilles heel of the flood silver finish is silver dendrites. In the presence of moisture and electrical bias, silver can form dendrites that can short-circuit adjacent leads of an IC. Some end users banned any silver outside the plastic encapsulant and this led to silver spot plated frames. It is notable that for flood plated finishes, either gold or silver, the solderable finish was applied at the lead frame maker. The assembly test (A/T) site put the IC together and shipped it with no further plating required.

It is also noteworthy that with the change from flood plated silver to spot plated silver finishing, the base metal changed from Fe/Ni to high Cu alloys, predominantly CDA194. The base metal changes were driven by cost, the Fe/Ni alloys are expensive, and more important, the higher thermal and electrical conductivity of the Cu alloys. The CTE mismatch problem between silicon and copper was mitigated by the shrinking size of the silicon chip.

9.3.3   Silver Spot Plating

In silver spot plating, the silver is confined inside the plastic of the IC. Spot silver provides a surface for gold wire bonding and eliminates the risk of silver dendrites. Base metal cleaning and silver spot plating are performed by the lead frame manufacturer. The IC assembly process heavily oxidizes the un-plated copper external leads, because of the thermal exposure at die attach and wire bonding. For silver spot plated lead frames, the external leads are plated after molding. This second plating step that includes oxide removal and mold flash removal is done in the A/T site and historically has been tin/lead (Sn/Pb) plating. With the need for lead-free in green designs, most A/T sites using tin/lead plating have converted to matte tin (Sn) plating. This process, comprised of plating steps at the lead frame maker and the A/T site, is one of the two IC finishing systems in widespread use today and is considered in detail in a later section.

Prior to surface mount technology (SMT) packages, many IC leads were finished by solder dipping. The dual in-line packages (DIPs) were relatively simple to solder dip given the relatively wide lead spacing on the finished packages. With the advent of finer-pitch gull wing leads for surface mount, solder dipping became less attractive because of solder bridging. There is little, if any, solder dipping done currently for termination finishing. Fine-pitch leads led to either post mold solder plating or preplated finishes (PPFs).

9.3.4 Nickel/Palladium (NiPd) and Nickel/Palladium/Gold (NiPdAu) Finishes

The NiPd and NiPdAu finishes are preplated finishes (PPF), where the entire lead frame plating operation is done at the lead frame manufacturer and there is no plating done at the A/T site. These are normally flood or full plating processes, involving no masking. This allows relatively simple and robust plating line design with very high throughput. Extra attention is needed at the lead frame manufacturer and the A/T to maintain finish integrity because the solderable finish that the end users (PWB assembler) see is produced by the lead frame manufacturer. There is no opportunity to "save" or rework an assembled IC with marginal solderability by replating it after molding. This process is the other IC finishing system in widespread use today, and is considered in greater detail in a later section.

In the IC component lead finish historical background, as shown in Fig. 9.1, there are currently two broad categories of lead-free IC component finishes: silver spot plating with a tin-based finish on the external leads and a PPF (non-tin-based) finish. The tin-based finishes are post mold plated, in the same fashion as tin/lead finishes that have been around for more than 30 years in the IC industry. This requires a plating operation in the IC A/T site.

The PPF typically use a layered structure of nickel and precious metals, for example, palladium and/or gold. The PPF are applied during the lead frame manufacture, not at the A/T site. This is the reason they are termed "preplated finishes". The PPF require no change to comply with the lead-free requirement. The structure, manufacturing process, and advantages and disadvantages of both of these lead-free finish systems are discussed in detail next.

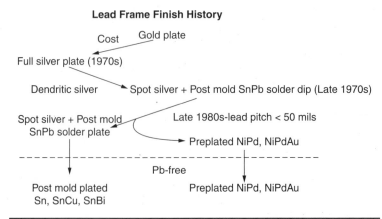

Lead Frame Finish History

Cost   Gold plate

Full silver plate (1970s)

Dendritic silver   Spot silver + Post mold SnPb solder dip (Late 1970s)

Spot silver + Post mold SnPb solder plate   Late 1980s-lead pitch < 50 mils

Preplated NiPd, NiPdAu

Pb-free

Post mold plated Sn, SnCu, SnBi   Preplated NiPd, NiPdAu

Figure 9.1 Lead-free component finish historical background.

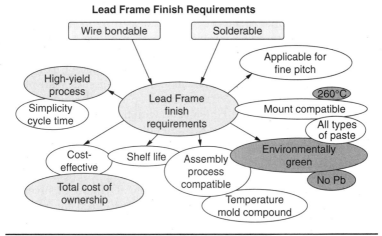

FIGURE 9.2 Lead frame finish requirements.

9.4   Component Finish Requirements

To understand the lead-free finishes, it is well to consider the minimum component substrate finish requirements or more specifically lead frame finish requirements, as shown in Fig. 9.2.

Of the many requirements shown in Fig. 9.2, the two primary ones are:

1. The finish should be wire-bondable to allow connection of the chip to the lead frame.

2. The finish should provide a solderable termination after IC assembly.

Additional requirements are properties that promote mold compound and die attach adhesive adhesion, passing moisture sensitivity tests, compatibility with both lead-free and tin/lead solders, extended shelf life, and economic viability.

9.5   Tin-Based Finishes for IC Components

The finish on the external leads of ICs is the prime concern of the PWB board assembly shop (the IC end user). For most lead-free tin-based finishes, this would be matte tin that is plated onto the exposed leads after the IC device is molded. Matte tin replaced the traditional tin/lead based post mold plated finishes in many applications.

Plated tin/copper (SnCu) and tin/silver (SnAg) have been suggested as possible lead-free termination finishes. However, there are two fundamental issues with such binary plating systems. First is the control of the alloy, in terms of percent tin versus percent of the non-tin metal (copper or silver). Second, measuring the thickness of binary coatings presents problems because when using x-ray fluorescence one must know the alloy composition. In some select applications, tin/bismuth (SnBi) has been used in lieu of matte tin. This finish poses additional cross-contamination issues with other component finishes at the PWB assembly shop. For more information, see Sec. 3.4.2 on tin/bismuth interaction with tin/lead.

Tin/lead alloys, being close to the eutectic, do not exhibit the binary alloy issues mentioned in the previous paragraph. Control of alloy content is relatively simple by using soluble anodes of the desired composition, that is, 60/40 tin/lead. In tin/copper and tin/silver alloys, the percentage of the non-tin metal (copper or silver) is relatively small, typically less than 5 percent, to keep the melting point of the alloy in a workable range for PWB soldering.

For tin-based finishes, there are complex IC assembly process flow steps that the A/T site must perform before molding, as shown in Table 9.1. The A/T site must first be able to attach the die (silicon chip) and then make wire bonds from the chip to the lead frame . For a lead frame used in a post mold plating process, the internal lead tips, where the gold wire bond is made, and normally the die pad are silver-plated using a selective plating process.

From the flow shown in Fig. 9.1, it can be seen that the lead frame finishing (plating) is done in two locations: the lead frame manufacturer silver spot plates the lead frame , and the A/T site plates a tin-based finish after the device is molded.

9.5.1   The Lead Frame Manufacturers' Plating Process Flow

Leadframes are manufactured in three steps: stamping (or etching), plating, and offset and cutting. The stamping (or etching) and the

Ag spot plated lead frame from lead frame manufacturer
Dispense die attachment epoxy
Place IC (chip)
Cure epoxy
Au wire bond from chip pads to lead frame internal leads
Overmold wire-bonded chip that is on the lead frames
Degate and deflash mold compound
Post **plate exposed metal** of the molded lead frame strip with matte Sn
Trim and form the external leads
Electrical test

TABLE 9.1   IC Assembly Process Flow with Post Mold Plating

offset and cutting processes are essentially identical for both tin-based and PPF. The plating steps for either post mold finished lead frames (tin-based) or PPF lead frames are different. In this section, the plating of tin-based finishes is discussed.

For the lead frame manufacturer, the silver spot plating process is well established with a nearly 30-year history. It requires masking of the lead frame by either a reusable rubber mask in a clamping mechanism or, for extremely tight spot location tolerance, a photoimageable plating resist. The mask or resist defines the area to be spot plated. The silver spot plating process flow is shown in Table 9.2.

For most high-volume lead frame applications, the lead frames are manufactured in continuous coils that facilitate stamping and plating. These coils are have the die pads offset and are cut into strips prior to shipment to the A/T site where there are handled in strip form. In the case of the reusable rubber mask, the plating line movement is "step and repeat." That is, a length of lead frame material is indexed, stopped, clamped, and plated, and then the cycle is repeated.

The photoimageable plating resist is not widely used as it involves cleaning the stamped lead frame, laminating, exposing, and developing the plating resist prior to plating. This is done in a separate wet processing line. The advantages of using a photoimageable plating resist are that the plating line can be run continuously and a resist plating mask produces excellent spot location.

The step and repeat process has limited throughput as it involves complex and time-consuming mechanical masking, opening and closing a mask, and indexing precisely a length of lead frame . It has yield and quality implications for mislocated silver spots and silver "bleed-out." In the latter case, either the reusable rubber mask is damaged, or the back pad doesn't seal the masked areas, allowing small amounts of silver to be deposited outside the spot area. If the bleed-out extends beyond the plastic body of the IC, silver dendrite growth can occur in the presence of humidity and voltage bias.

There is also the issue of backstripping. This is reverse plating of the lead frame after silver spot plating to clean up any autocatalytically deposited silver from the copper base metal of the lead frame. It can also eliminate minor silver bleed-out, but at a cost. The backstrip process entails overplating the silver spot thickness so that after backstripping, the minimum silver thickness is obtained. Typically silver spot plating is 1.5 to 3.0 μm thick.

As noted earlier most lead frames are now built on copper-based alloys, typically CDA 194, there is also the need to preserve the very active copper surface that results from silver spot plating and backstripping. This can be done by an antitarnish agent, such as benzotriazole (BTA). BTA can be applied from solution or by sublimation from BTA-impregnated wrapping paper. Nevertheless, silver spot plated on copper base metal lead frames are highly susceptible to tarnish and contamination, either airborne or from handling.

Payoff
Splicing station
Clean—alkaline soak
Rinse/contact
Electroclean—alkaline
Rinse/contact
Activate—H_2SO_4 or $H_2SO_4 + H_2O_2$
Rinse/contact
Cu strike—thin layer of Cu to provide pure Cu surface, not base metal alloy
Rinse/contact
Ag strike—to provide readily "strippable" surface for backstrip
Rinse/contact
Drive system
(*All prior steps are with continuously moving lead frame strip*).
Accumulator (slack loop for step and repeat indexer to draw from)
Ag spot plating head—pneumatically actuated clamp with silicone rubber mask and back pad
Step and repeat driver—typically mechanical gripper on air piston to index lead frame strip a discrete length
Rinse
Accumulator (slack loop to take indexed lead frame material)
All subsequent steps continuous
Rinse/contact
Backstrip
Rinse/contact
Multiple rinses
Drying system
Drive system
Takeup

TABLE 9.2   Ag Spot Plating Process Flow [for continuous coil plating (reels of lead frames about 300 m long)]

Finally, there is the part changeover process in the production line. It is time-consuming and involves changing plating masks and spargers, which are the anodically charged nozzles behind the mask that supply electrolyte, as well as adjusting the feed length of the indexer, to accommodate the wide variation in lead frame pitch and length.

All the preceding process steps are completed at the lead frame manufacturing site. The output of this site to the A/T site consists of wire-bondable lead frame strips with an offset die pad to accommodate the thickness of the chip and a surface that can accept die attach epoxy and mold compound with good adhesion.

9.5.2 The Assembly/Test Plating Process Flow

As noted above, for a silver spot plated lead frame for tin-based finishes, the A/T site mounts the chip, performs the wire bonding, and overmolds the chip with plastic, leaving the external leads exposed for soldering to a PWB. The external leads are trimmed and formed to electrically isolate them and provide a shape for PWB mounting, such as gull wing, through hole, and J-lead. Electrical test is performed after trim and form.

The external lead plating takes place between molding and trim and form, since the entire lead frame/chip/plastic assembly is electrically continuous, and is shown in Table 9.3. After trim and

Loader—picking station that loads individual strips to a continuous clamping belt
Cam station—opens clamp on belt to accept strip
The belt never stops moving
Loader must articulate in *xyz* dimension to match belt speed *z*
All electrical contact provided by clamping belt
Clamping belt is usually titanium
Deflash—chemistry designed to dissolve plastic
Clean—alkaline soak
Rinse
Electroclean—alkaline
Rinse
Activate—H_2SO_4 or $H_2SO_4 + H_2O_2$
Multiple matte Sn plating stations, depending on desired speed
Rinse
Multiple rinses
Drying system
Drive system
Unloader
Belt stripping—on backside of line, electrochemical stripping of clamping belt to remove plated Sn that is on tips of clamps

TABLE **9.3** Post Mold Plating Process Flow (for strips of molded lead frames with wire-bonded ICs inside the plastic)

form, the leads are electrically isolated. The plating task is to take a roughly 20-cm-long strip of heavily oxidized copper alloy, with small blocks of thermoplastic mold compound and associated mold flash embedded in it, and plate it with a lead-free finish that will provide a solderable surface with some degree of shelf life. Handling the strips of lead frames is more complex than feeding a continuous coil of lead frames into a step-and-repeat plating line. This type of plating line is costly, has a large footprint, and has limited throughput because of the loading/unloading process.

The preceding process is well defined for tin/lead and matte tin; however, the matte tin chemistry is still undergoing improvements for tin whisker mitigation and to extend solderability shelf life. This type of strip-to-strip plating line is commercially available from several manufacturers. As such, the operational pitfalls of strip-to-strip matte tin plating are arguably well defined.

Various mitigation techniques have been proposed to slow down tin whisker growth. Unfortunately, a key technique is to provide a nickel underplate prior to matte tin plating. This added plating step was not planned for in the tin/lead strip-to-strip plating lines, and consequently the A/T is faced with a Hobson's choice: plate the nickel in a separate, new dedicated line, or install a nickel plating station prior to the matte tin station in existing lines. Either choice is expensive. Separate plating lines affect yield and cycle time and entail significantly more floor space. Adding another plating station to a legacy line entails lengthening the line (if space is available) to insert the additional plating cell.

A key to successful strip-to-strip plating is loader/unloader setup. The distance between successive strips must be short and reproducible, on the order of 2 mm. This minimizes "dogboning," where the high current density at the end of the strip results in rapid metal deposition, and thick plating. Maintaining tight and reproducible strip-to-strip distance in effect "reassembles" the continuous coil for which there are only two ends.

Running a strip-to-strip plating operation in an A/T site involves placing a wet chemical process inside a dry electromechanical environment. It is a costly and labor-intensive process, involving specialized engineering and management resources not required for IC assembly. The facility's support for the plating line includes process water treatment (reverse osmosis/deionized), chemical waste treatment, air handling exhaust and scrubbers, analytical laboratory with wet chemistry, and ideally surface analysis capability with SEM/EDX and specialized quality control equipment to assess solderability and tin thickness (X-ray fluorescence).

After plating, there is the trim and form operation that segregates the termination leads and forms them to end product shape. There are potential problems from mechanical damage to the matte tin plate since tin is a soft metal, with a plated thickness on the order of 10 μm.

The trim and form tooling can gouge the tin and create burrs that flake off and lodge randomly on the finished product. However, the largest risk with tin plating is tin whiskers. This was addressed in Chap. 3 and is discussed in a later section in this chapter.

9.6   PPF Component Finishes

As noted in the prior section, PWB board assembly shops worry about the finish on the external leads and their solderability. What's inside the plastic is the concern and responsibility of the IC supplier. For PPFs (non-tin-based finishes), the external leads are plated with nickel that is protected from oxidation by some overlying set of precious metals, typically palladium and/or gold.

The A/T site attaches the die (silicon chip) and makes wire bonds from the chip to the lead frame. The die attachment and wire bond performance for PPFs must be functionally transparent to the A/T site, that is, no degradation in yield or throughput when compared to Ag spot plated lead frames. The post mold plating process is not required for PPFs, since the entire lead frame is plated with nickel plus precious metals at the lead frame supplier. The nickel plus precious metals provides a solderable surface after IC assembly at the A/T site. The IC assembly process flow with PPFs is shown in Table 9.4. The internal lead tips, where the gold wire bond is made, the die pad, and all the rest of the lead frame including carrier rails that are trimmed away in trim and form are nickel, palladium, and/or gold plated using a flood or overall plating process.

For this flow, lead frame finishing (plating) is done in one location at the lead frame manufacturer. As a consequence, there are no wet processes in the A/T site and no supporting engineering or wet process controls, and no waste treatment is required.

PPF lead frame is provided by lead frame supplier
Dispense die attachment epoxy
Place IC (chip)
Cure epoxy
Au wire bond from chip pads to lead frame internal leads
Overmold wire-bonded chip that is on the lead frames
Degate and deflash mold compound
Trim and form the external leads
Electrical test

TABLE **9.4**   IC Assembly Process Flow with PPF Lead Frames

9.6.1   The Lead Frame Manufacturers' Plating Process Flow

As noted earlier, lead frames are manufactured in three steps: stamping (or etching), plating, and offset & cutting. The stamping (or etching) and the offset & cutting processes are essentially identical for both tin-based and PPF. For the lead frame manufacturer, the PPF process is well established with a nearly 20-year history. It requires no masking of the lead frame, since it is a flood plating process. A step-and-repeat plating line is not required. For most high-volume lead frame applications, the lead frames are manufactured in continuous coils that facilitate stamping and plating. These coils are offset and cut into strips prior to shipment to the A/T site, the same as for silver spot plate/post mold plate ICs. The PPF process flow is shown in Table 9.5.

Payoff
Splicing station
Clean—alkaline soak
Rinse/contact
Electroclean—alkaline
Rinse/contact
Activate—H_2SO_4 or $H_2SO_4 + H_2O_2$
Rinse/contact
Ni strike—thin layer of NiCu to present pure Ni surface, not base metal alloy to Ni plate baths
Rinse/contact
Ni plating baths—multiple baths to plate sufficiently thick Ni to prevent Cu diffusion. More plating baths or cells equal higher throughput
Rinse/contact
Pd plating—short plating time. Pd is plated very thin
Rinse
Rinse/contact
Au plating—shorter plating time, Au is plated ultrathin
Rinse/contact
Multiple rinses
Drying system
Drive system
Takeup

TABLE 9.5   PPF or Non-Tin-Based Plating Process Flow [for continuous coil plating (reels of lead frames about 300 m long]

The functions and thicknesses of each layer in the nickel palladium gold version of the PPF are as follows:

1. Nickel 0.5 to 2.0 μm
   a. Barrier layer to prevent Cu diffusion
   b. Provides an Au wire-bondable surface
   c. Is solderable by either SnPb or Pb-free solders
2. Palladium 0.01 to 0.10 μm
   a. Prevents oxidation of the Ni surface
   b. Rapid dissolution during soldering
3. Gold 0.003 to 0.009 μm
   a. Allows reduction of required Pd thickness
   b. Rapid dissolution during soldering

For the nickel/palladium version of the PPF, the minimum palladium thickness would be increased to 0.075 μm.

The nickel/palladium and nickel/palladium/gold plating lines run extremely fast, up to three times faster than silver spot plating lines. This is an advantage in capital utilization, space, personnel, and other throughput-related factors. However, the high speed does pose process control issues. The plating lines run so fast that in the time it takes to measure the thickness of nickel, palladium, and gold, even with dedicated modern X-ray fluorescence instruments, an entire coil (300 m) of lead frame material can be processed. The need for real-time parametric control of the plating line is a result of the high speed, thin deposits, costly raw materials (palladium and gold), and quality requirements of the next-step customers: the A/T site and then the PWB assembly shop.

Most users of nickel/palladium and nickel/palladium/gold based finish require real-time process monitoring with automatic shutdown of out-of-control processes within the plating line. At a minimum, the line speed and nickel, palladium, and gold plating currents must be constantly monitored as time and amperage determine plating thickness.

9.7 Comparison of PPF and Silver Spot/Matte Tin Plating

Figure 9.3 is a schematic comparison between a nickel/palladium/gold finished IC and a silver spot plated/matte tin post mold plated IC. On the right-hand side is the silver spot plated frame with relatively thick silver inside the plastic mold compound. Outside the plastic is the very thick matte tin plating, which would have been tin/lead prior to the lead-free era.

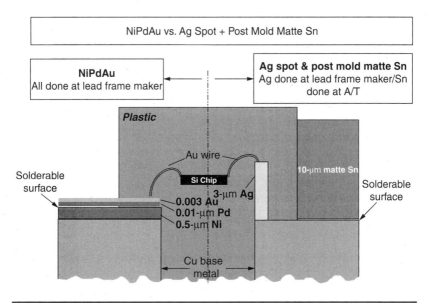

FIGURE **9.3**   Comparison between a NiPdAu finished IC and an Ag spot plated/matte tin post mold plated IC.

The cost of the nickel/palladium/gold plated PPF lead frames is typically more than silver spot plated lead frames. The intrinsic value of the gold and palladium in the lead frame is much greater than the value of the silver in the silver spot plated lead frame, using historical silver, palladium, and gold prices. But, looking at total cost of ownership, one must factor in the elimination of the post mold plating step when using a PPF. And, there are savings in capital equipment, yield losses, cycle time, and operational costs that offset the higher cost of PPF lead frames relative to silver spot plated lead frames.

The scrap from trim and form with matte tin plating is low-grade copper scrap, netting a discounted value for the copper or no value. Nickel/palladium/gold plated trim and form scrap, despite the minute thickness of palladium and gold (~10 and 2 nm, respectively) is class 1 copper scrap, meaning the scrap settlement is on high-grade copper, plus palladium and gold value. Normally no value is returned for the nickel. It is important to keep track of the scrap when using PPF and use it to offset the cost of the lead frames.

PPFs have no risk of whiskers, which is the foremost advantage. At trim and form, since the nickel, palladium, and gold are quite thin and hard relative to tin, the propensity for burr and flake formation is eliminated in a well-maintained process.

Soldering and the mechanism for soldering are quite different for the two systems. Considering only the external leads (outside the plastic), for the matte tin finished or any other tin-based plating, the

Metal	μm/s, 215°C	μm/s, 250°C
Ni	<0.0005	0.005
Pd	0.0175	0.07
Cu	0.08	0.1325
Au	1.675	4.175

TABLE **9.6** Dissolution Rates in Tin/Lead

soldering mechanism is fusion or melting of the termination finish. Reflow temperatures for lead-free solders are typically in the range of 235 to 260°C, comfortably above the melting point of tin (232°C). This is contrasted with the melting points of palladium and gold which are far higher than an IC can withstand (Pd 1552°C, Au 1064°C). The palladium and gold *dissolve* into the molten solder during reflow. Thus, it is extremely important that the thickness of the palladium and gold be held thin and consistent. The dissolution rates of nickel, copper, palladium, and gold are shown in Table 9.6. It can be seen that the dissolution rates of both palladium and gold far exceed that of the nickel.

For PPF, the solder joint is made to the nickel layer. The palladium and gold dissolve into the solder paste and are undetectable after reflow in the bulk solder joint. For matte tin, the solder joint is made to either copper or nickel, depending on how the matte tin is plated, whether a nickel underlayer is put down prior to matte tin plating.

9.8   Tin Whiskers

A lot of research has been done and publications made on tin whiskers, particularly since the lead-free requirement has driven the electronics industry to consider matte tin plating for solderable terminations. A detailed discussion of this topic is given in Chap. 3. In addition, publications by iNEMI are particularly useful and comprehensive on this subject, and there is an excellent website from NASA on tin whiskers.

Testing for tin whiskers is a major operational issue for IC suppliers, aside from reliability issues. The JEDEC tin whisker test procedures are time-consuming and costly. The test procedures offer only a "snapshot" of the matte tin plating process. Because tin whiskers can take decades to develop, the tin whisker tests can be viewed at best as indications of tin whisker propensity. An example of a tin whisker grown under JEDEC specified conditions is shown in Fig. 9.4.

As noted previously, there are "mitigation" techniques that can be used, in some cases mandated by end users, to reduce the propensity

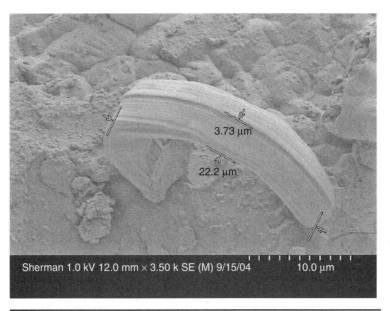

3.73 µm

22.2 µm

Sherman 1.0 kV 12.0 mm × 3.50 k SE (M) 9/15/04   10.0 µm

FIGURE **9.4** Typical tin whisker grown from plated matte tin deposit.

for tin to whisker. None are considered foolproof. Nickel underplate, cited earlier, prevents copper diffusion into the tin grain boundaries and thus precludes the formation of copper/tin (CuSn) intermetallics. It is thought by some that the (CuSn) intermetallics in the grain boundaries generate compressive stress in the deposit, and the tin whisker growth is a stress relief mechanism. Matte tin itself was thought to be less prone to whiskers than so-called bright tin because matte tin has larger grains (fewer grain boundaries), but recent data dispute this claim. Reflow of the plated tin or "annealing" of the tin has also been thought to be another tin whisker mitigation option. The annealing is normally required within 24 h of matte tin plating. All these mitigation techniques add cost and operational control issues.

The seriousness of the tin whisker problem is evidenced by the exemption from the RoHS lead-free legislation of "mission-critical" applications: aerospace, medical, military, and telecommunications. The irony of the situation is that while most of the industry is laboring to get lead out of the product for compliance with RoHS legislation and testing for lead in components and finishes, the exempt sectors of the industry are testing to be sure that lead is in the finishes, so there is no risk of tin whiskers and for robust solder joints.

The other part of the reason for exemptions is the meager reliability and long-term usage record of lead-free solder joints that use the tin/silver/copper (SAC) alloys. The mechanical and aging behavior of the SAC solder is poorly understood with no good models in place at this time to predict failure. For lead-free electronics, the fact is that any application in which a failure can result in a loss of life or in catastrophic consequences for society is exempt.

9.9 Moisture Sensitivity Level

Moisture sensitivity level (MSL) performance is a secondary effect of moving to lead-free for the IC manufacturer and ultimately the PWB assembler. The higher reflow temperatures required by lead-free SAC solders impose more severe demands on plastic IC packages. This in turn leads to degradation in MSL performance. The JEDEC MSL J-STD-20D is shown in Table 7.1 moisture sensitivity levels from Chap. 7.

There are several ways to improve MSL. One approach is to alter either the mold compound (encapsulant) or the die attach adhesive or both. This presupposes consistent lead frame surface for bonding. Changes to the adhesives and the plastics are beyond the scope of this chapter. Enhancing adhesion of the lead frame surface *with any and all* mold compounds and adhesives is the main issue. It is important to strive for a universal solution since the lead frame manufacturer has little control over which plastics and adhesives will be used in the A/T site that uses the lead frames.

PPF in general has an advantage over silver spot plated copper since the PPF surface is very well defined, consistent, and by design resistant to oxidation. Silver spot plating can exhibit a problem with the "bare" copper surface inside the plastic, outside the silver spot area. The chemical state of the copper can vary widely because of the highly active copper surface that results after silver backstripping. There can be different levels of copper oxidation depending on plating conditions such as:

1. Rinse water pH
2. Temperature of final drying boxes on plating line
3. Whether an antitarnish agent is used
4. Degree of backstripping
5. Residual immersion (autocatalytically deposited) silver on the surface
6. Whether a copper strike was used prior to silver spot plating
7. Whether a silver strike was used prior to silver spot plating

The result of these factors is that the copper surface to which the mold compound and die attach adhesive adhere can vary, among lead frame suppliers or among lots from a single supplier. There is also the impact of A/T conditions; the thermal exposures of die attach and wire bonding can be different at different A/T's. All these factors can lead to a suboptimal surface for adhesion. Nevertheless, adhesion to copper with controlled oxidation generally is superior, despite the pitfalls of silver spot plated frames.

For PPF, the surface is well defined. It is typically 3 to 9 nm of gold over palladium. The oxidation state is not an issue. The gold is not oxidized since by design the finish is resistant to oxidation. The surface is consistent from different suppliers and lots within each supplier, again by requiring well-defined and controlled processes. The issue with nickel/palladium/gold or nickel/palladium finishes is that by dint of having no oxidation, there is no polar functionality on the surface for the mold compound or die attach adhesive to latch onto.

For tin/lead PWB board soldering processes with reflow temperatures of 217°C, achieving MSL = 1 performance with either silver spot plated or nickel/palladium/gold plated lead frames is typical. Converting to the lead-free reflow domain of 235 to 260°C initially led to an MSL loss of 1 to 2 classes. Refer to Sec. 7.5.1 for more information about MSL.

Currently, for both lead-free finishes, die attach adhesion and mold compound adhesion improvement have recaptured the MSL performance for the majority of ICs. There is still room for improvement in specific areas, for example, thermally enhanced packages with either exposed die pads or heat slugs. Finishing options for improved adhesion are currently under intense investigation; see Chap. 7 for more information on MSL.

9.10   Summary

The conversion to lead-free electronics has driven profound and largely un-forecasted changes in the semiconductor manufacturing process and materials infrastructure. The wholesale shift from tin/lead solder plating to matte tin plating, with the signal exception of PPFs, has opened potentially high risks associated with tin whiskers.

The higher reflow temperatures of the lead-free process place greater stress on the IC chip itself and the IC package. The MSL drop seen initially with lead-free conversion is evidence of the extra stress on the package. The IC makers and their material suppliers have responded with improved lead frame surfaces and higher-performance die adhesives and mold compounds.

The end users, exempt from the lead ban, such as military, space, and medical, are leery of commercial off-the-shelf ICs being finished with tin/lead and the risk of inadvertently introducing matte tin, whisker-prone parts into their products. A cottage industry has cropped up converting matte tin finished leads to a tin/lead finish. In some instances there has been the unneeded conversion of nickel/palladium/gold to tin/lead. The pitfalls with converting matte tin to tin/lead are the extra thermal and chemical exposure (stripping agents and flux) to which the IC is subjected. Not surprisingly, conversion processes may void any warranty of the IC manufacturer.

The PPF option for lead frame finishing and IC assembly eliminates many of the concerns that a lead-free process creates: no whiskers, reduced thermal and chemical exposure of the finished IC package, and compatibility with both lead and lead-free processing.

Acknowledgments

The chapter author would like to acknowledge Oscar Nietzel, retired from Texas Instruments, for his enthusiastic and patient guidance on lead frames and plating for electronics. He would also like to thank Doug Romm of Texas Instruments for numerous discussions and efforts associated with development of the NiPdAu finish.

References and Websites

Abbott D., et al. "Palladium as a Lead Finish for Surface Mount Integrated Circuit Packages." *IEEE Transactions CHMT*, 14: 567 (1991).
Bader, W. G. "Dissolution of Au, Silver Pd, Copper and Ni in a Molten Tin/Lead Solder." *Welding Research Supplement*, pp. 551–557, December 1969.
Glazer J., P. Kramer, and J. Morris. "Effect of AU on the Reliability of Fine Pitch Surface Mount Solder Joints." *Proceedings of the Technical Program, Surface Mount International Conference & Exhibition*, pp. 629–639, August 1991.
http://ec.europa.eu/environment/waste/weee/index_en.htm
http://www.inemi.org/cms/
http://nepp.nasa.gov/whisker/
http://www.jedec.org/

CHAPTER 10

Nanotechnology Opportunities in Green Electronics

Alan Rae, Ph.D.
NanoDynamics Incorporated, Buffalo, New York

10.1 Introduction

Nanomaterials and structures are normally characterized by having at least one dimension less than 100 nm (10^{-9} m). There are a wide range of materials and structures that can be produced by a wide range of techniques, hence ISO's decision to call its TC229 committee "Nanotechnologies."

Small size—around the wavelength of light—can produce interesting properties. Below the wavelength of light, nanostructures can become invisible to the naked eye; band gaps in semiconductor materials can be modified to alter electrical and optical properties; metals can sinter and coalesce well below their melting temperatures; and nanotubes and nanowires can behave as individual transistors.

How green are nanomaterials? Most are made from high-yield "bottom-up" processes such as precipitation from solution or vapor rather than energy-intensive grinding and milling processes. Some have been used for many years, for example, "submicrometer" silica fillers, carbon black, precipitated barium titanate, colloidal antimony oxide, or CMP materials. In fact, any integrated circuit with features less than 100 nm might be classified as nano!

How about the safety issue? Nanomaterials, just as regular chemicals (<10 nm) or particulate materials (>1000 nm) do, will vary in toxicity. With growing regulation of chemical species particularly in

Europe evidenced by the RoHS and REACH regulations, there is an awareness that we need to be professional in how we commercialize these materials. Recent U.S. publications include "Environmental, Health, and Safety Research Needs for Engineered Nanoscale Materials" (National Nanotechnology Initiative, September 2006, www.nano.gov) and "Progress Towards Safe Nanotechnology in the Workplace" (NIOSH Nanotechnology Research Center, February 2007, www.niosh.gov).

This chapter explores the different types of materials that can be used in electronics, how they are made, the effect they can have on processing electronic products, and their environmental impact.

The following topics are discussed:

- Why nanotechnologies are important to green electronics
- When and where nanoelectronics will appear in the workplace
- The manufacture of nanomaterials
- Application areas in electronics as a whole
- Nanoelectronics as an enabler for additive processes such as printed electronics
- Nanoelectronics as an enabler for low-temperature solder assembly

10.2   Why Nanotechnologies Are Important to Green Electronics

Nanotechnology is receiving a lot of attention from companies, universities, and governments. The U.S. National Nanotechnology Initiative, currently running at an investment level of over $1.3 billion per year, is matched by initiatives in Europe and Asia, and this level of funding is exceeded by private industry funding for a total worldwide effort of approximately $10 billion in 2007. But what does it mean for existing businesses and new businesses in the electronics market? Is it a real tool for today, or are the applications way out in the future? Will it be economic or outrageously costly? Will we solve problems or create new ones?

In reality, nanotechnology is a tool kit for the electronics industry; it gives us tools that allow us to make nanomaterials with special properties modified by ultrafine particle size, crystallinity, structure, or surfaces. These will become commercially important only when they give us a cost and performance advantage over existing products or allow us to create new products.

The collection of synthesis techniques collectively known as nanotechnology presents many opportunities to reshape the electronics industry from top to bottom. Nanotechnology can offer us

- Uniform particles—metal, oxide, ceramics, composite
- Reactive particles, as above
- Unusual optical thermal and electronic properties—phosphors
- Nanostructure materials—tubes, balls, hooks, surfaces
- Self-assembly—liquid-based, vapor-based, or even by diffusion in the solid state

10.3 When and Where Nanoelectronics Will Appear in the Workplace

The iNEMI road map (www.inemi.org) is a comprehensive survey that reviews the issues affecting the electronics supply chain. Gaps in the technology or infrastructure that can adversely affect iNEMI members are identified, and the iNEMI Research Committee was formed to prioritize and disposition the tasks and identify companies, universities, and government laboratories that can address them for the mutual good.

Almost every road map chapter identifies aspects of nanotechnology that can enhance existing products or replace their structure or function.

1. *Long-term opportunities.* Once CMOS technology dips below about 20-nm resolution, quantum effects such as electron tunneling start to result in phenomena such as unacceptable leakage; the only way to move below that size is to utilize quantum effects in new types of minute structures, be they pure electronic or bioelectronic (remember, the most effective and energy efficient computer available sits on your shoulders). We know that if we extrapolate Moore's law, we hit the wall with CMOS about 2015, and although we don't know which technology will disrupt it, it may well be disruptive.

2. *Middle-term opportunities.* In many areas of technology, once we hit an area of concern, we can develop a work-around. Hence clock speed, which was the measure many used as the measure of processing capability, has been replaced by distributed processing with two processors placed on the same chip. This gets the job done without the extreme heat penalty and gives us a breathing space—many upper-end processors generate between 100 and 200 W—but the heat issue has not gone away. It basically depends on the number of transistors switching and the efficiency of each transistor, which is being modified with the use of high-dielectric-constant dielectric at nanothickness levels. Several unusual properties of nanoscale

materials—enhanced thermal conductivity of carbon nano-tubes, diamondlike films, nanometal dispersions—have the promise of aiding heat removal.

We are accustomed to seeing Energy Star stickers on domestic appliances outlining their absolute and relative power efficiency but not yet on many consumer and electrical devices. There has been a strong emphasis on improving batteries and reducing power consumption through better displays, more efficient processors, etc., and 8-hour laptop battery life is being projected for 2008 (see, for example, the Intel initiative described at http://onlinetoolkit.intel.com/docs/busi/ExtendingBatteryLifeWP.pdf). We can now expect this to extend to cell phones and other battery-powered activities and beyond this to energy-using devices such as power supplies which are generally purchased for cost rather than energy efficiency.

3. *Shorter-term opportunities.* Enhancement of shielding materials, solders, conductive adhesives, underfills, etc. is now becoming possible as nanosize materials become available and economic.

10.3.1 Technology Push versus Market Pull in Nanoelectronics Deployment

Many nanomaterials have been developed because of their interesting properties, and companies have been founded on products for which there is limited market demand (see the left half of Fig. 10.1).

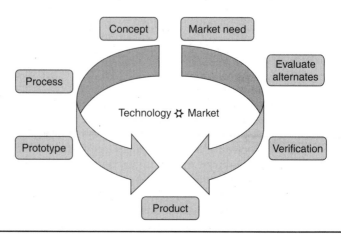

FIGURE 10.1 The balance between technology push and market pull.

This tends to produce leading-edge products with very limited immediate commercial potential. Work by iNEMI and others suggests the time for deployment in the electronics industry is typically 7 years for a new product that fits with the existing infrastructure and 15 years for a disruptive product. Doubters only need to look at the intensive phase of lead-free solder qualification and implementation (1999 to 2007—and still not complete for complex boards) or the implementation of micro electro-mechanical systems (MEMS) devises in accelerometer applications—30 years! The electronics industry is fast-moving in terms of ultimate product development but very conservative when it comes to accepting new materials, devices, and systems.

Companies taking the technology push approach are very vulnerable in this industry type and often have run out of funding before revenues materialize.

The faster approach may be the market pull approach where existing solutions are sought for market needs. This conservative approach can result in a very small increment in performance which in the end may not show a cost-benefit improvement for that particular application. Thus, a conservatively run company may be outflanked by a smaller, more progressive competitor. It can also mean that a technical solution is implemented when not all the supply chain issues have been worked out. We have seen this many times in the introduction of novel consumer electronics where quality and supply issues have led to shortages and recalls. This balance is illustrated in Fig. 10.1.

A more balanced approach being taken by a number of companies is to take a parallel track, constantly reviewing technology choices on a portfolio basis and applying them to market needs. Technology platforms thus developed, such as printable electronic materials, diamond-like coatings, carbon nanotubes, or nanosolders, can be applied to several other business areas in addition to pure electronics, such as structural engineering, life sciences, or energy.

Nanotechnologies should be applied only where there is an economic advantage coupled to a performance advantage. This is seen in industrial processes as well as consumer goods where a "luxury" new technology becomes the standard once the existing technology is reaching its limits and the new technology starts to take hold. An example in the consumer field is photographic film versus VCR versus DVD versus DVD-R versus DVR versus live feed software transfer; in the industrial field, disk capacitors versus multilayer capacitors versus embedded capacitors versus integral capacitors. In each case the shape of the graph may differ, but the important point is that there is a crossover which marks the start of market adoption of the new product. What is tough to predict is the "tipping point" when new products will become dominant. This is illustrated by Fig. 10.2.

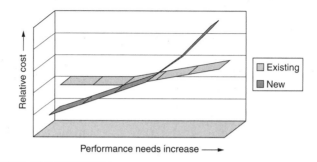

FIGURE **10.2** Replacement of an existing technology by a new technology.

10.4 Manufacture of Nanomaterials

Manufacture of nanomaterials can be top-down or subtractive (milling, grinding, melting, and spraying) or alternatively bottom-up or additive (precipitation, deposition, vapor processing). In general, the top-down processes are more economic at larger particle sizes above 1 μm but bring high energy cost and poor yield below 1 m. Bottom-up processes because of their additive nature are the lower-cost route to particles less than 100 nm but tend to be less economic above 1000 nm (1 μm).

Figure 10.3 outlines the process required to precipitate metal nanopowders, and Fig. 10.4 outlines the range of processes that can

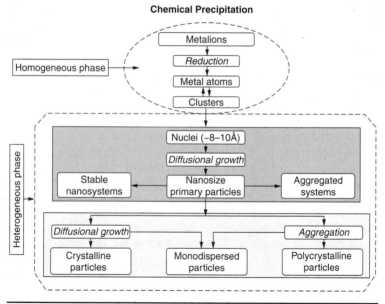

FIGURE **10.3** Precipitation sequence for nanometals. (Courtesy of Dr. D. V. Goia, Clarkson University.)

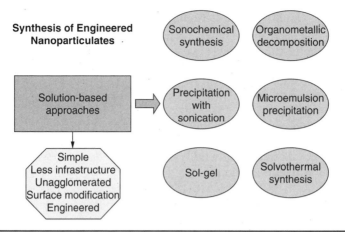

be used to modify the precipitation approach to develop particular sizes and configurations of coatings.

Above and beyond the solution systems there exists a whole range of vapor phase techniques that can be used to build up particles and coatings. Later in this chapter we will see the application of atomic layer deposition of alumina at nanometer thickness to alter the chemical properties of materials such as the thermal interface material filler boron nitride. We will also discuss carbon nanotubes which are produced using vapor phase carbon sources and solid/molten catalysts under a wide range of conditions.

10.5   Application Areas in Electronics

10.5.1   Semiconductor

Some of the most revolutionary applications in nanotechnology are in the semiconductor areas. As the semiconductor road maps look out toward 2015 and below 20-nm features, the need for different structures is becoming apparent—once we move to ultraviolet and then X-ray lithography, there is nowhere to go (in a practical sense) to image ultrasmall features.

Imagine doping a carbon or silicon nanotube, coating it with differently doped materials, and assembling it (preferably self-assembling it) in an array. Imagine creating quantum dots that can store a single electron charge. Imagine trapping atoms inside a nanotube and using the electron spin to create a quantum computing device. There are a large number of potential routes to new computing, storage, and optical devices. The devices we are making now are quite clumsy compared with established semiconductor technology. But they will improve!

10.5.2 Electronics Packaging

Packaging of advanced devices is going to continue to be problematic since temperature is the enemy of ultrafine features which can be easily destroyed by thermal diffusion. In the shorter term, improved fillers for mold compounds (which are already ~90 percent filler) can lead to better thermal and electrical performance combined with easier flow properties.

Thermal management will continue to be a major issue, and the use of high-thermal-conductivity carbon nanotubes and diamondlike films with thermal conductivity over twice that of copper will provide worthwhile solutions.

10.5.3 Board and Substrate

In the last 5 years, the board business has changed from a commodity business to a specialty business with materials optimized for thermal, high-frequency, and environmental reasons. The only problem is that commodity prices are still demanded! Boards still need improvement in areas such as CTE and flatness, and the embedding of passive components needs a low-cost self-assembly type of process to lower the costs to make it truly competitive.

Improved ceramic substrates are possible, but in fact many ceramic operations use the principles of nanotechnology already in developing precursor particles with high reactivity and uniformity. There are options to improve conductor and embedded passive technology and to strengthen the substrates as average substrate size is increasing due to the greater use of modules.

10.5.4 Passive Components

There are many uses of interconnection materials within passive components. Monosize materials hold promise in tantalum capacitors and ceramic capacitors, where improved termination materials and electrode materials promise a further reduction in size and cost. In particular, reducing the electrode thickness in base metal ceramic capacitors promises a significant improvement in volumetric efficiency. Some of the metal powders that can be used are 200-nm Ni powder for internal electrodes and 600-nm Cu powder for termination pastes, as shown in Figs. 10.5 and 10.6.

Advanced copper and nickel nanopowders offer the following advantages:

- Controllable particle size
- Virtually monosize particle size distribution
- Monolayer oxidation-resistant coating
- Economic and scalable process

5 μm

FIGURE 10.5 A 200-nm Ni powder for internal electrodes. (Courtesy of NanoDynamics Inc.)

5 μm

FIGURE 10.6 A 600-nm Cu powder for termination pastes. (Courtesy of NanoDynamics Inc.)

Barium titanate and other dielectric materials are already produced by "wet processes" such as hydrothermal and oxalate precipitation that produce nanosize barium titanate particles that are subsequently grown by thermal treatment to develop optimum properties. As we move down beyond dielectric thicknesses of 1 μm, we may have to move to technologies such as gel casting or vapor-phase technologies to optimize performance.

Again, the passive components system is under extreme price pressure and cannot afford to improve performance by using advanced materials. Indeed in some quarters we have seen a rolling back of technology to meet pricing demands which does not bode well for the future of the industry.

10.5.5   Display

Display technology is similar to semiconductor technology, where the number of techniques to assemble devices is bewildering! There are enhanced plasma displays using nanotubes, light-emitting nanowires, and high-output quantum dot or phosphor arrays. A lot of development work needs to be done, but new products can be expected in the next 5 years. Although carbon nanotubes have been extremely expensive, at thousands of dollars per gram, new processes are dramatically reducing the cost of production and the cost will reduce past the cost of silver flake within 5 years.

10.5.6   Device Shielding

Nanotechnology can also contribute in case design and EMI shielding. A major cell phone manufacturer has stated in public that they will produce nanotube-reinforced phone cases that are stronger, have lower weight and EMI shielding, and are recyclable. Nanomaterials can also make device cases more attractive and wear-resistant and can even impart antibacterial properties. Nano-oxides in surface finishes can more than double the hardness and scratch resistance of transparent clear coats or pigmented coatings. Self-cleaning UV-activated coatings as well as antimicrobial coatings are in use today.

10.5.7   Power Delivery

A great deal of work is being done in this area to increase the properties of portable energy systems, by enhancing the performance of lithium-ion batteries and fuel cells.

Nanosize electrolyte, anode, and cathode materials used in solid oxide fuel cells have already increased the performance of solid oxide fuel cells (SOFCs) while reducing the operating temperatures and start-up time dramatically. These products will become commercially available in 2008, such as the Compact SOFC using nanotechnology manufactured by NanoDynamics Inc (see www.nanodynamics.com).

10.6 Nanoapplication Examples

10.6.1 Conductive Adhesives

Most conventional isotropic (conductive) adhesives use silver flake in a resin base, typically epoxy, and have limitations in terms of conductivity, strength, and moisture resistance. They typically cure in the 120 to 150°C range, much lower than the reflow temperature of solders.

Copper flakes with oxidation resistance and thickness of 150 nm, shown in Fig. 10.7, and silver coatings are also being employed in conductive adhesive and EMI shielding.

Anisotropic adhesives are largely based on Au plated polymer spheres dispersed in a resin. As the resin shrinks on cooling, the conductive surfaces are pulled together and the spheres compressed between them.

There are many opportunities to use novel conductive adhesive systems to reduce processing temperatures, but issues of strength under physical shock and high-moisture environments still have to be resolved.

Novel attachment opportunities such as conductive hook and loop and other types of mechanical fasteners based on carbon nanotubes have the potential of replacing conventional interconnections as strong entangled loops promise thermal and electrical conductivity, compliance, and strength, calculated at up to 30 times the strength of

9 μm

Figure 10.7 Oxidation-resistant copper flake, of thickness 150 nm. (Courtesy of NanoDynamics Inc.)

conventional adhesives on an area basis (see news@nature.com, October 22, 2003). Arrays of hooks can be synthesized which can entangle other hooks or a nanotube mat. This type of biomimetic assembly is being studied by iNEMI (http://www.inemi.org/cms/projects/ba/Mechanical_Assembly.html) and has the potential to revolutionize the value chain and economics of electronics assembly.

10.6.2 Nanotechnology as an Enabler for Additive Processes such as Printed Electronics

The way we form circuits has essentially remained unchanged for many years. Circuit boards are formed from copper laminated onto epoxy-glass and then selectively etched, laminated, drilled, and plated. Integrated circuits are attached by wire bonds or solder bumps. Devices are connected using solder paste screen printed onto the circuit board and then melted in a reflow oven.

Margins were squeezed dry in the circuit board industry by the collapse of electronic markets in 2001, consolidation of customers, the development of significant low-cost capacity in Asia, global infrastructure threatening regional markets, brutal competition, and an increasing demand for specialty products to cope with lead-free and high-temperature requirements, but at commodity prices.

The subsequent recovery has not improved margins as the strong demand for commodity metals such as copper and tin from Asian demand and the rising cost of oil-based polymers have eroded margins further. The limited ability to pass on increasing raw material and energy costs has put further pressure on a beleaguered industry that finds it increasingly difficult to make existing products profitably and finds it even more difficult to make the investment to develop new ones.

Meantime, outside the mainstream electronics industry, printing is being explored as a way of developing circuits by start-up and established companies supported by government and private funds worldwide.

The concept of printing circuit traces is not new. In fact the technique has been used in ceramic hybrid circuits and in flexible circuits used in membrane switches and keypads for many years. The printed electronics market is difficult to quantify exactly right now because definitions differ, but what is agreed by many experts is that it is poised to grow dramatically over the next 5 years.

What is driving this change? In fact it is a combination of new materials, new circuit structures, and new market opportunities. Many of the markets are nascent, the structures are not optimized, and the materials still require further development; but all areas are receiving great attention worldwide with a potential of at least $10 billion by 2010 (www.nanomarkets.com).

Printing Technologies

Printing techniques are of interest for a number of reasons.

1. *Environmental.* Printing processes are additive. Many circuit forming processes are additive-subtractive, and typically it can take up to 8 kg of material to produce a 1-kg circuit board which builds in cost constraints as well as environmental ones. This is so because of a significant number of plate-deposit-expose-develop-etch-strip-wash types of processes that involve large quantities of chemicals and water at each step.

2. *Flexibility.* It can be digitally driven serial deposition, or it can be massively parallel deposition using flexographic or lithographic printing. Materials can be deposited on three-dimensional surfaces such as casings by using ink jet or transfer printing. Digital gives flexibility, parallel gives low cost.

3. *Cost.* It can be adapted to low-cost processes, for example, reel to reel on flexible substrates. For the last several years, the fastest-growing sector of the substrates business has been in flexible substrates—traditionally polyimide for solderability, but polyester is of course used widely as a low-cost substrate in keyboards and membrane switches.

4. *Low-temperature processing.* Nonfired composite curing below 200°C or low-temperature fired systems below 250°C can be used to create functional circuit elements.

Materials which can be printed include the following:

- *Conductors.* To use low-cost substrates such as polyester and even paper instead of epoxy, polyimide, or ceramic, process temperatures must be reduced to below 200°C.

- *Semiconductors.* Polymers or polymer composites can be printed as components of structures such as solar cells (Graetzl cells), light-emitting diodes for displays, or transistors.

- *Dielectrics.* These include high-K, for example, for embedded capacitors, or low K for insulation.

- *Phosphors and other functional materials*

Competing processes include these:

- Plating processes. These are well established but use aggressive chemical baths.

- Etching laminated planar copper on glass-epoxy to develop traces and pads. This is a well-established low-cost mature technology.

- Semiconductor processes, e.g., spin-on, lithographic, chemical vapor deposition (CVD), atomic layer deposition (ALD), etc.

Traditional printing normally does not require as high a level of precision in registration as for electronics. Our eyes are very forgiving when we are reading the printed word or viewing pictures. Newspapers, for instance, are frequently printed with misregistrations approaching nearly 1 mm. Sophisticated screen printers used in the printing of solder paste on circuit boards on a single-panel basis run to much closer tolerances and contain extremely complex optical registration and measurement systems; see examples of these advanced printer types at www.speedlinetechnologies.com or www.dek.com.

According to the iNEMI 2007 road map (www.inemi.org), commercial boards typically use a 75-μm (3-mil) line width and space with state-of-the-art microvia substrates running at 50-μm line and space. The tendency toward finer resolution for smaller components is balanced by the move to higher clock speeds and wireless systems where larger conductors are needed to optimize impedance at high frequencies.

It is considered that 50 μm is just about the limit for stencil printing where the minimum stencil thickness to allow reasonable life is also around 50 μm. Conventional color lithographic or flexographic prints normally run with a registration of 400 μm but can be "tightened" somewhat for electronics use, and dot sizes for ink jet can allow a resolution of 50-μm line and space with very thin conductors. Xerographic and electrophoretic printing processes have already been demonstrated with compatible accuracy and precision.

The resolution required will depend on the application. Printed RFID antennas on cardboard packaging or keyboard networks require much less precision than trimmed electrodes.

Conductive ink fillers are particulate, and for ink jet applications they are screened to eliminate agglomerates or particles larger than 1 μm, or 1000 nm. In general, smaller is better, and an average particle size of 50 to 500 nm seems to be the accepted range. Other printing techniques with lower resolution can tolerate larger particles.

The Ink Jet Process

Ink jet processes are only part of a wide range of printing tools available to the industry including flexographic, lithographic, stencil, etc. Different techniques will be appropriate depending on the speed, precision, and thickness of deposition desired.

So the "sweet spot" for data-driven printing can be described as follows:

- Products that cannot be readily miniaturized to the nanoscale, e.g., keyboards, antennas.
- Short-run products, e.g., prototypes.
- Thin and three-dimensional structures. Note that repeated print applications can increase thickness at the expense of speed.

- Trimmable resistors, capacitors, and inductors potentially with a metrology feedback loop to allow tunability to the desired value (a major issue with passives where engineers are used to the tight tolerances obtained by testing and sorting every discrete component).

The primary data-driven technology is ink jet which is in widespread industrial use for product marking and in the office and home environment for low-cost color printing. Electrophoretic or xerographic processes also have the potential to be data-driven, but are less advanced in commercialization than ink jet.

There are two main types of ink jet, continuous or drop on demand. In the continuous process, a stream of ink droplets, produced usually by a piezo device creating pressure pulses through an orifice, is ejected and subsequently caught in a collection cup. The stream can be deflected electrostatically, at which point droplets reach the substrate. These jets are frequently used for high-speed marking of rapidly moving food packaging and generally have lower dot size precision (~50 μm) than drop on demand.

Drop on demand ink jet systems can use a piezo or a thermal engine. Electrostatic and acoustic as well as electrostrictive engines are also possible. Generally piezo systems are lower-resolution (~125 μm) than thermal engines (~20 μm) but are often favored in industrial ink jet systems. Variable drop size metering can increase precision, and multiple nozzles enhance throughput.

Conductive Inkjet Technology (CIT) (see www.xennia.com/conductive_inkjet/) has developed a laser-assisted drop on demand technology to lay down fine lines down to 5 μm, creating essentially invisible conductive lines. The highest precision so far appears to have been achieved at the University of Cambridge (http://www.technologyreview.com/InfoTech/wtr_14676,294,p2.html) where 100-nm features were developed in organic polymer circuits. Both of these techniques rely on the deposition of organic rather than purely metallic structures—in the case of CIT the polymer is subsequently metallized on a catalyst, and in the case of Cambridge, a conductive polymer is used.

Ink jet heads are attached to reservoirs which may be in the form of sealed and refillable cartridges, similar to consumer ink jet cartridges, or may be custom-designed with features such as recirculating pumps to keep solids suspended. Water-based inks are typically formulated to a viscosity of about 10 cP (0.01 Pa·s, about the viscosity of blood). The diameter of the ink ejected from the orifice is usually (but not always) about the same as the orifice diameter. Typical orifice sizes are in the tens to hundreds of micrometers and are manufactured by photolithographic or laser ablation techniques. For a very complete description of ink jet structures, see H. P. Le's excellent tutorial at www.imaging.org.

Marking inks can be hot melt, solvent-based or water-based. Water-based or hot melt inks are generally preferred for environmental and safety reasons, but solvent systems tend to be more flexible. Pigment contents are typically 5 to 10 percent, but for functional systems the solids content needs to be pushed as high as possible; 50 percent is a target which is not readily achievable.

At high concentrations of very small particles, the particle-to-particle distance becomes very small, and it is difficult to prevent agglomeration. There are 1 million times more particles in a 10-nm suspension than in a 1-μm suspension of the same mass! To put this in another perspective, solder pastes typically contain about 90 percent by weight (50 percent by volume) of metals. Increasing the surfactant content can increase the loading at the expense of burn-out difficulties at a layer stage, as many surfactants contain significant inorganic moieties.

Smaller particles are less susceptible to Stokes' law settling (even at 80 nm under an optical microscope, significant movement through brownian motion is visible!) but need to be prevented from agglomeration through management of the solvation sheath and need to be prevented from oxidation or other undesirable reaction with the environment.

Silver concentration is a typical compromise as silver, a typical conductor former, with a specific gravity of 10.5, tends to settle quickly at high concentrations at the low viscosities needed for fine ink jetting according to Stokes' law, unless the particle size is less than 50 nm. A typical water-based ink viscosity, as mentioned above, is about 10 cP, similar to blood, whereas the viscosity of solder paste at rest is similar to that of peanut butter, or 150,000 cP.

Ink Materials for Conductors

The choice of conductor pigment is determined by the resistivity required and the process parameters. Although conductive oxides such as indium tin oxide and other nonmetal pigments such as carbon blacks can be used, their relatively low conductivity means that they are really resistors rather than conductors. There is no easy mechanism to coalesce the particles, and so percolation conduction is the only practical mechanism. Carbon nanotubes promise higher conductivity but are still relatively expensive and difficult to disperse and align into useful conductor structures.

In the metals area, precious metals such as gold are practical but not economic; low-melting metals are scarce or toxic (mercury, gallium, indium), oxidize readily, or do not consolidate at practical processing temperatures to form high-conductivity films.

Nanosilver ink has received attention because of its potential to be made in stable suspensions that can be ink-jetted to form deposits that can be made conductive by processing below 200°C.

Many materials exhibit a change in properties as they move toward the nanoscale. This is due to the increase in the relative proportion of higher-energy surface atoms. This change can be exhibited as a change in reactivity, for example, sinterability or electromagnetic properties driven by band gap changes resulting in dramatic changes of electronic properties or optical properties such as color and transparency.

Where these changes occur, the tipping point is a function of the individual element or compound and its environment and normally occurs below 100 nm. In the case of silver it is exhibited as a dramatic increase in sinterability at low temperatures below 200°C.

The tipping point in properties found in certain materials at the nanoscale level is shown in Fig. 10.8.

Nanosilver powders can be produced in essentially monodisperse form using, for example, the precipitation technology developed at Clarkson University. They are precipitated using a bottom-up approach where nuclei grow under a protective polymer coat that allows metal atoms to accumulate while acting as a charge director to keep the embryo crystals separated. The crystals are allowed to grow until the desired particle size is achieved; then the reaction is stopped, and the crystals are suspended in an appropriate vehicle or dried. The polymer coating, which can be made hydrophilic or hydrophobic, aids redispersion and prevents the spontaneous sintering of the dried silver particles. Figure 10.9 shows a 10-nm Ag dispersion (left) and 60-nm dried Ag powder (right).

Additional developments include the precipitation of nanosize Ag platelets which can be made compatible with ink jetting through a narrow orifice (compare with regular silver flake in conducting adhesives which is typically 10 μm in diameter). A general rule is that in ink jet inks the maximum pigment particle size should be less than one-tenth of the orifice diameter to avoid bridging and blockage. The ratio may need to be greater depending on the nozzle profile and operational cycles.

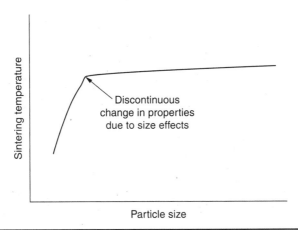

FIGURE 10.8   The "tipping point" in properties found in certain materials at the nanoscale.

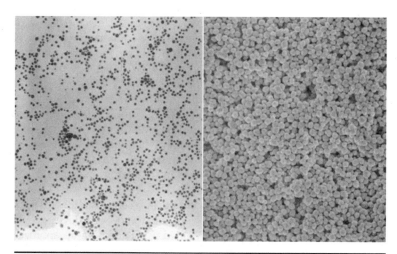

FIGURE **10.9**   A 10-nm Ag dispersion (left) and 60-nm dried Ag powder (right). (Courtesy of NanoDynamics Inc.)

Inks can be prepared by adding the particles to a suitable vehicle, applying them to the substrate, and then consolidating them in a polymer matrix or by sintering particles together at a temperature compatible with substrate integrity. Significant silver-to-silver conductivity is achieved, and sintering occurs when the polymer coating is removed at 125°C or above; or if the polymer is a permanent matrix rather than a process binder, percolation paths are ensured by the very large number of nanoparticles in the composite. Note that this process temperature is well below the traditional melting temperature of 961°C for silver, and the remarkable silver mobility can be explained by the energy of particles with a high surface-to-volume ratio—10,000 times higher at 10 nm compared with 1 μm.

The processing temperature can be matched to the substrate by modifying the silver size and surface preparation as well as the ink formulation.

Nanosilver can be dispersed in alcohol for compatibility with styrene-acrylic resins or in glycols for compatibility with water-based systems. Curing is by cooling of hot melt resins, evaporation of solvents, heat crosslinking, or UV crosslinking. Scanning laser techniques, IR lamps, and other methods of selectively applying heat have been suggested.

10.6.3   Nanoelectronics as an Enabler for Low-Temperature Board Assembly and Thermal Interfaces

Low-Temperature Assembly Drivers

The move to environmentally preferred materials (that is, the restriction of hazardous materials in manufacturing) and more ecology-friendly products through market pressure and legislation necessitates the need for higher-melting-point no-lead solder alloys for electronic components and printed-wiring board interconnects. These higher-melting-point alloys result in higher processing temperatures. Currently, nonlead solder alloys such as "SAC," or Sn/Ag/Cu, have a temperature above liquidus in the range of 217°C and higher. The higher-melting-point solder composition results in higher reflow process temperatures well above the liquidus, typically above 240°C. The higher reflow temperatures can negatively affect product reliability due to higher residual stresses in PWB assemblies; these temperatures require tougher qualification requirements for components and sometimes a significant change in the manufacturing processes.

Warm manufacturing is a term developed by iNEMI (http://www.inemi.org/cms/projects/ba/Warm_Assembly.ml) as a way to describe processes that can be used to assemble electronic devices at temperatures lower than that of solder reflow. The need for warm manufacturing is becoming more and more critical due to

- Increased reflow temperatures due to the conversion to lead-free solder causing failure of existing components and devices

- Increased thermal sensitivity of newer devices that can contain nanoscale semiconductor structures, MEMS devices, proteins, and other low-temperature organic materials

One example of the use of nanotechnology to modify assembly processing temperatures is to use the excellent work that has been developed over the past several years on melting point depression. The phenomenon of melting point depression of nanoscale metal particles has been studied since the 1960s, when it was noticed that extremely thin evaporated particles of metal have a lower melting point than does the bulk material.

Developing Low-Temperature Assembly Systems

We have experienced this type of temperature-sensitive assembly issue once before in the electronics industry in the late 1990s when optical communications became mainstream and integrating optoelectronic components onto boards became a real headache. Surface-mountable packages were not available due to the materials of construction including low-melting alloys, the need for optical alignment within the package, and the temperature sensitivity of the optical fiber "pigtail," typically 80°C.

In the case of the optoelectronics industry, the infrastructure drove the process, as tends to happen with many new developments, over a period of about 10 years.

- The device packages were lead-attached and hand-soldered post reflow to boards in a slow and cumbersome process. Pigtails were manually coiled, fiber was manually spliced, connectors were manually cleaned, and the assemblies were clipped to the boards.

- Automation of manual processes used robotics to duplicate the human processes followed.

- The next step was to develop package formats standardized by multisource agreements (MSAs) some of which had solderable bases and pluggable optical packages or used high-standoff wave-solderable leaded packages.

- A limited number of surface mount reflowable packages became available.

We can see a similar scenario playing out with the next generation of temperature-sensitive electronics components—a plethora of package types, different levels of thermal sensitivity, and even analogs of the optical pigtails in the form of optical sensing or "plumbing" to allow remote access to gases or liquids. Some of these will evolve in the way that optoelectronics parts did, to satisfy the infrastructure; but others may utilize some of the new technologies becoming available which will be complementary to solder processes—but at much lower temperatures.

A typical connection is a multifunction link which will conventionally provide a joint with the following properties (solder figures in parentheses):

- Physical connectivity (~ 4000 lb/in^2, 50 MPa)
- Ductility and compliance ($\sim$50 percent elongation)
- Thermal connectivity ($\sim$30 W·m/K)
- Thermal expansion ($\sim$30 ppm/°C)
- Electrical connectivity ($\sim$10 Ω·cm)

For a comprehensive compilation of properties of metals used in the electronics industry, see http://www.boulder.nist.gov/div853/lead%20free/props01.html.

The connection also must exhibit chemical resistance, adhere to contact surfaces, and have low toxicity (at least lower than tin/lead alloys).

Not every interconnect needs all these properties, but this is the type of interconnect that the electronics infrastructure expects. Are these properties adequate in every case? No. For example, some solder joints have to be assisted by underfill to meet drop-shock and flexural fatigue requirements for handheld devices.

Low-Temperature Assembly Options

Use conventional solder assembly with lead-free solder and do the following:

- Make the device more rugged. We are starting to see this happening in optoelectronic systems where components can now be fabricated directly onto silicon rather than assembled in an optical train of discrete lenses and components. This approach is viable when volumes justify the nonrecurring engineering (NRE) expenses and the devices can be fabricated in this way, but many of the newer devices depend on thin films or wires or may have attached enzymes or other active species that cannot be incorporated readily. This will be available in several years.

- Insulate the package during reflow. Systems are available to provide temporary local cooling to packages during reflow and rework. See, for example, http://www.alphametals.com/products/thermal/ which can be tailored to protect package contents while allowing reflow of solder bumps on the bottom of the package. This is available now.

- Insulate the package. This approach has an obvious size and footprint penalty and is currently available.

- Use a soldered socket with lock-in or snap-in contacts. This approach works well with the existing infrastructure but also carries a size (especially height) and footprint penalty and is currently available.

Use a conventional solder assembly and a second, lower-temperature pass with the following:

- *Selective soldering.* Laser or other techniques can be used to locally heat laterally spread contacts without heating the package. This is an approach that has grown greatly in popularity over the last few years. The penalty is lateral spread of the package footprint and the time penalty due to the fact that this type of soldering is normally a serial application of heat rather than a parallel application.

- *Low-temperature alloy solder.* There are a number of lower-melting alloys such as tin/bismuth and a range of indium alloys which melt between 100 and 200°C. Unfortunately, the cost of indium alloys is high due to the demand for indium in indium/tin oxide transparent conductors for displays, and many bismuth alloys exhibit low ductility. In addition these metals, if operated above a homologous temperature of 90 percent (the homologous temperature is the ratio of the ambient temperature to the melting temperature in degrees absolute), exhibit creep properties which reduce the joint strength significantly. For example, an alloy melting at 150°C (423 K) would start to soften significantly at 90 percent of 423 K = 381 K (115°C), which is well within the operating range of many electronic assemblies.

- *Amalgams.* Amalgams present an intriguing opportunity to use a liquid metal and a solid metal intermixed to form a paste which then forms a relatively high-melting amalgam as in the dental fillings which many of us carry. This amalgam is about 50 percent mercury with the balance being an alloy of tin/silver and copper! Unfortunately the only liquid metals available at room temperature are conductive adhesives. A range of adhesive systems can be cured at temperatures as low as 80°C. In general, conductive adhesives exhibit good electrical conductivity in the range of 0.001 to 10 Ω·cm and relatively low thermal conductivities around 1 W·m/K. Compositions optimized for electrical conductivity may not show the best thermal conductivity and vice versa. Cure times range from 3 to 60 min. Conductive polymers tend to have lower conductivities and charge carrier mobilities; filled systems tend to use silver or other fillers or flakes at a high loading >80 percent. Highly filled systems tend to have flexural strengths lower than that of solder and are not suited for many handheld applications.

Nanosolder Solutions

Solder materials containing nanosize metals exploit the high surface area and high surface energy (think of it as stored energy) of nanosize particles to lower the apparent melting point below the conventional melting point. Thus, the melting points of tin, silver, and copper, the ingredients of lead-free solder, can all be depressed below 200°C, well below the eutectic melting point of 217°C. Putting these materials in a printable solder paste is the goal of an iNEMI project which is characterizing the metals and is looking to develop a proof of concept demonstration, a model similar to the tin/silver/copper SAC system that was developed for lead-free solders (http://www.inemi.org/cms/projects/ba/Pb-free_nano-solder.html).

Many materials exhibit a change in properties as they move toward the nanoscale. This is due to the increase in the relative proportion of higher-energy surface atoms. This change can be exhibited as a change in reactivity, for example, sinterability or electromagnetic properties driven by band gap changes resulting in dramatic changes of electronic properties or optical properties such as color and transparency. Where these changes occur, the tipping point is a function of the individual element or compound and its environment and normally occurs below 100 nm. In the case of silver it is exhibited as a dramatic increase in sinterability at temperatures below 200°C, as described earlier in the chapter.

Nanosolders will be available within the next 2 to 3 years. Sinterable silver systems are available now.

"Wet" Adhesives

As described earlier, enhanced adhesives using nanomaterials can enhance mechanical properties (even a 0.1 percent addition of multiwall carbon nanotubes can raise the flexural strength of an unfilled epoxy by 30 percent) as well as electrical properties due to the large number of potential contact or tunneling events when nanosize particles are present. A reduction of 10 times in particle size (for example, from 1 µm to 100 nm with the same weight content yields a $10 \times 10 \times 10 = 1000$-fold increase in the number of particles present.

A number of iNEMI members are participating in a SPIR project at the University of Binghamton (http://watson.binghamton.edu/level2/industry.html#SPIR) to quantify the effects of nanoparticles—metal and carbon-based—in resin-based systems to get a consistent data set to characterize performance.

"Dry" Adhesives

As described earlier, these are biomimetic nanoattachment processes. Hook and loop fasteners (http://www.velcro.com/) used the ideas generated by plant burrs to create a whole new fastening paradigm. A further extension would be to use carbon nanotubes as suggested by researchers at Michigan State University in 2003. Further information can be found at http://www.pa.msu.edu/~tomanek/tomanek.html.

Carbon nanotubes can be made in curved shapes (indeed it is rather difficult to make them straight!) that could be used to develop room-temperature interconnects. Textured dry adhesives based on the gecko foot approach (biomimetics), where nanosize hairs attach to surface roughness using van der Waals forces, have been the subject of a great deal of research; see, for example, http://www.uakron.edu/news/articles/uamain_1293.php. Now these can be duplicated using carbon nanotubes at a number of research locations with strengths far higher than those that comfortably attach a gecko to your ceiling!

Challenges to Implementation

The addition of fine particles to improve the conductivity of polymer systems has been well documented in the literature. However, duplicating the performance in real life tends to be more difficult. A sample of the finest nanosilver that is "thrown over the wall" is almost certain to fail because of a number of factors listed below. If these factors are addressed, then it will be possible to achieve product improvements.

Virtually all nanometals need to have a protective coating that allows dispersion in liquids and prevents oxidation or self-sintering—uncoated silver will sinter to itself at room temperature, and uncoated copper will turn black as you watch. Uncoated aluminum and many other metals are pyrophoric. This is an advantage in high-temperature sinterable coatings such as frit-based multilayer ceramic capacitor (MLCC) termination systems, but the coatings that work in that application may not work in lower-temperature or adhesive systems. Coatings may be applied in vapor phase, by precipitation or by chemical reaction, for example, ALD alumina coatings on BN or metals.

Agglomeration is another concern. Nanomaterials have high surface energy and love to stick to one another. Because the number of particles per unit volume is high, the interparticle distance is lowered as the particle size decreases for a given content. At the level under 5 μm, many colloids agglomerate at concentrations of only 3 to 5 percent.

Nanoparticles supplied as dry powders are notoriously difficult to disperse in viscous liquids. The tendency is now to supply them as dispersions in compatible liquids or polymer master batches.

Segregation may be a problem if it is inadvertent but can be an advantage if it is directed toward self-assembly. By varying the hydrophobic and hydrophilic nature of the surface y coating, it is possible to encourage the material to disperse in a similar medium or to preferentially congregate on a second phase or at the surface.

Figure 10.10 shows a self-assembled 30-nm silver nanoparticle monolayer on a polymer surface through hydrophobic surface modification of a normally hydrophilic system.

None   SEI   7.0 kV   X150.000   100 mm   WD 7.7 mm

FIGURE 10.10 Self-assembled 30-nm Ag nanoparticle monolayer on a polymer surface. (NanoDynamics Life Sciences Inc.)

Reaction with catalysts, fillers, or other constituents is a major consideration in process design; nanoparticles by their nature are reactive. That's why we are using them! Mixing a borderline pyrophoric metal powder with organic peroxide is not something you should try at home!

Many of the same issues of surface reactivity apply to sinterable systems (e.g., ink-jettable silver). Control of surface is key to controlling reactivity, but the same compromises apply as in adhesives; particles have to be disagglomerated, remain in suspension, and achieve a practical concentration and shelf life. The issue is compounded by the fact that many printing systems need relatively low viscosities (tens of centipoise) between the viscosity of water and milk, and many of the fillers we need to disperse have high specific gravity compared with the organic pigments we use in many applications—silver, for instance, has a specific gravity of over 11.

The main issue in nanosolder production is tin. It is a very reactive metal and loves to oxidize (most of the dross forming on tin/lead solder is actually tin oxide). Take it down below 50 nm and there can be real issues. It is, however, possible to produce tin particles in a reducing environment, and in the conference presentation we will update progress on the microstructure and melting behavior of tin and tin-based solders being investigated under the iNEMI nanosolder program.

There are several challenges in producing dry nanotube adhesives: producing the material in the correct form, verifying the theoretical strength and conductivity parameters, designing device interconnects, and verifying processability and reliability—and doing so economically! Nevertheless, this is a "game-changing" route to assembly that could have real benefits for our industry.

Thermal management has been highlighted by the ITRS and iNEMI electronics road maps as a key impediment to developing better-faster-smaller-cheaper electronic products. The more transistors used, the more heat is evolved. The situation will become worse when devices approach the size where quantum effects take hold and leakage currents rise.

Thermal interface materials (TIM) and heat spreaders can be enhanced in a number of ways by nanotechnology. Carbon nanotubes have received a lot of attention because of their extremely high thermal conductivity along the longitudinal axis. But to dissipate heat effectively, they have to be surface-treated to induce better coupling to a matrix (e.g., solder); aligned together by flow, electrophoretic, or other post processes; or even grown in array structures on a suitable substrate. Technologies also exist to cross-link them to form a two-dimensional mat.

With non-electrically-conducting TIM one of the major materials in use is hexagonal boron nitride (BN). It's a wonderful material with the same structure as graphite and equivalent thermal conductivity

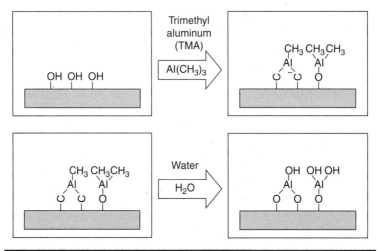

Figure 10.11   Sequential alumina ALD coating. (Courtesy of ALD NanoSolutions Inc.)

but is an effective electrical insulator. The only problem with BN is that its hydrolyzed surface is essentially boric acid which interferes with many polymer matrices and limits the effective concentration that is allowable before viscosity becomes effective.

Atomic layer deposition adapts a technique known to the semiconductor industry for planar deposition and applies it to particulates using fluidized-bed technology. The availability of economic organometallic raw materials with appropriate processing temperatures determines the availability of coatings. Oxide coatings such as alumina, titania, or zirconia or metal coatings can be built up atomic layer by atomic layer to provide a thin coherent coating. In the case of boron nitride, trimethyl aluminum reacts with the hydroxides on the surface, releasing methane; subsequent hydrolysis attacks the remaining methyl groups; and layer by layer an alumina coating is built up. Wank et al. [*Powder Technology*, 142: 59–60 (2004)] showed that this ultrathin layer could dramatically reduce the viscosity of loaded epoxies and improve peel strength while maintaining a high thermal conductivity. Figure 10.11 shows a sequential alumina ALD coating.

This will be a specific niche application but a technologically valuable use of the nanotechnology with a high potential payoff for thermal interface adhesives.

10.7   Conclusions

Nanotechnologies offer a number of tools that will help produce electronic devices at lower temperatures and using additive processes that consume less material and energy than conventional processes.

Acknowledgments

I would like to thank colleagues at NanoDynamics Inc. and Metamateria Partners LLC for their contributions to this chapter in a practical way. Particularly I would like to mention Eric Groat, Gregg Berube, Sang Beom Lee, Dick Schorr, Suv Sengupta, and Rao Revur. Our university collaborators also deserve particular mention: Dan Goia and Clarkson and Carol Handwerker come to mind. Finally, I thank my colleagues at iNEMI in the Research and Technical Committees, and industrial partners, as well as coauthors and collaborators on several papers including Mark Chason and Andrew Skipor at Motorola. Some of the content of this chapter has been presented at various SMTA (Surface Mount Technology Association, Edina, Minnesota) meetings and is used with their permission.

References and Sources

References, many as websites, are embedded in the text or appended to the text as journal references and are current as of the date of writing.

For general reading on lead-free solder issues, the reader is referred to the author's chapter (Chap. 15) in the *Handbook of Area Array Packaging* (edited by K. Gilleo, McGraw-Hill, New York, 2002). Further information on the iNEMI lead-free nanosolder and warm manufacturing initiatives can be found at www.iNEMI.org.

Information on NanoDynamics Inc. and Metamateria Partners LLC can be found at www.nanodynamics.com. Contact information for Carol Handwerker and her group at Purdue University may be found at https://engineering.purdue.edu/Engr/People/ptProfile?resource_id=11509.

Information on ALD and ALD NanoSolutions may be found at www.aldnanosolutions.com. Information on printed electronics may be found on a wide range of company websites and articles, but the most useful primer the author has read is *Printed Organic and Molecular Electronics* (edited by D. Gamota, P. Brazis, K. Kalyanasundaram, and J. Zhang, Kluwer Academic Publishers, 2004).

Finally, an excellent source of information on a wide range of applications of nanotechnology and useful links is the National Nanotechnology Initiative website www.nano.gov.

Author Biographies

Chapter 3

David Pinsky is an Engineering Fellow with Raytheon Integrated Defense Systems. David chairs both the Raytheon RoHS Technical Steering Committee and the Raytheon Tin Whisker Core Team. He is a member of the Lead-Free in Aerospace Project—Working Group (LEAP-WG). He has 25 years of experience in reliability of soldered assemblies, materials selection for long-term reliability, and failure analysis of complex systems. He has published in the areas of tin whiskers, reliability of commercial components in military systems, hermeticity, solder properties, failure analysis techniques, and connector reliability. David received a B.S. in physics and an M.S. in materials engineering, both from the Massachusetts Institute of Technology.

Chapter 4

Ken Degan is the Enabling Technology Manager for Teradyne, Inc., in the operations organization. His responsibilies include advance technology development in the areas of attachment, mechanical, thermal, PCB, and test technology development to align with Teradyne's new products. He also led the Teradyne Environmental Strategy and Integration team. Ken joined Teradyne in 1984, holding positions in Test Technology and Concurrent Engineering. He holds a BSEE from the University of Massachusetts, Amherst, and a Certificate in Advanced Management from Babson College.

Chapter 5

Mark Quealy was introduced to the world of electronics manufacturing as a process engineer for defense and commercial applications while working for GTE Government Systems, General Dynamics, and Schneider Automation over the last 20 years. He is currently a Project Manager for Schneider Electric, a recognized leader in electrical distribution and automation and control. Mark began his RoHS journey investigating lead-free soldering and RoHS solutions with the New England Lead-free Consortium. Assigned to guide the implementation the RoHS directive for Schneider Electric's Connectivity Products End User Business unit in North Andover, Massachusetts, he quickly recognized the scope and difficulties of converting products to the new materials.

A graduate of the University of Massachusetts, Amherst and Lowell, Mark holds a Master's in project management from Boston University and is a certified Project Management Professional (PMP) since 2005.

Chapter 6

Robert Farrell

Robert Farrell has been the principal advanced development engineer for the Benchmark Electronics lead-free corporate team from 2002 to present. He is responsible for direct on-site support for a number of Benchmark Electronics manufacturing facilities as they convert to lead-free. He holds a Bachelor of Science degree in mechanical engineering from Union College and a Master's of Science degree in mechanical engineering from Worcester Polytechnic Institute.

Scott Mazur

Scott Mazur is the principal engineer and RoHS specialist for the Benchmark Electronics Hudson, New Hampshire, division. He is responsible to provide direct support for the customer base and the Hudson facility for the lead-free and RoHS manufacturing conversion. He has been a member of the Benchmark Electronics lead-free corporate team for the last two years. He holds a Bachelor's of Science degree in electrical engineering from Merrimack College.

Chapter 7

Richard A. Anderson, Ph.D., is a Senior Principal Engineer in Strategic Research and Development at Tyco Electronics, Wireless Systems Segment—M/A-COM, Inc. Dr. Anderson is responsible for package and substrate development including process development and component and assembly reliability. Current work includes area array (flip-chip, CSP, BGA) and small-size (QFN, MLP) millimeter wave processing and packaging, materials and process selection for volume production of nonhermetic RwoH (reliability without hermeticity), and microwave and millimeter wave assemblies for automotive, aerospace, and defense applications. As part of his development responsibilities, he helped form M/A-COM's Environmental Compliance Committee and serves as technical consultant. Dr. Anderson received a Ph.D. in materials engineering from Rensselaer Polytechnic Institute in electronic properties of materials. His prior work experience includes 3M, Foxboro Company, and Motorola. He holds eight current and pending patents in the electronics processing area.

Chapter 8

Michael J. Taylor has over 31 years' experience developing and launching new products and processes including keyboards, flexible circuits, semiconductor packaging, materials, solid-state sensors, and security devices, as well as 13 years' experience with technical start-up companies. Currently, Mr. Taylor works in Research and Development for Dynamic Details Inc. (DDi). Mr. Taylor was Director of Process and Product Development for MSC Electronic Materials and Devices and held technical leadership positions at Sheldahl, Flexible Products, Inc., Rogers Corporation, and Thiokol. Mr. Taylor has a strong materials and process background in circuit fabrication and assembly (roll to roll and panel), including additive and subtractive technology, adhesive-based laminates and cover films, cast polyimide, and substrates created with vacuum and laser fabrication techniques. Mr. Taylor is an experienced practitioner of statistical process control and modern quality systems ISO and TS, systematic troubleshooting processes, operator-friendly documentation, and technology transfer. He holds seven issued patents and has eight more in current applications. The patents are in multilayer printed-wiring boards, switches, memory components, optical materials, flexible circuits, and conductive ink. Mr. Taylor has a B.A. in chemistry from the University of Arizona, Tucson.

Chapter 9

Don Abbott is a Texas Instruments Fellow and Technical Programs Manager in semiconductor packaging. He received his B.A. from Bowdoin College and his Ph.D. from Northeastern University in analytical chemistry with emphasis on high-performance liquid chromatography. He has worked in the fields of electrodeposition, electronic packaging, and lead frame manufacturing for the last 29 years. He holds more than 40 patents.

Chapter 10

Dr. Alan Rae is the Vice President of Innovations, NanoDynamics Inc. Rae brings to NanoDynamics many years of experience in technology commercialization and international business. While with Cookson Group PLC, he was instrumental in ensuring the success of start-up and developed businesses including structural ceramics, flame retardants, refractories, electronic ceramics, and wafer plating systems. From 1999 to 2004 he was Vice President of Technology for Cookson Electronics, deeply involved in developing business opportunities for materials systems and equipment in silicon wafer fabrication, packaging, circuit board manufacture, circuit board assembly, and recycling. He is Director of Research for iNEMI Inc. (the International Electronics Manufacturing Initiative), a member and past Chair of the JISSO North America Committee (facilitating electronics business worldwide by harmonizing technology road maps and standards), and a member of the Industrial Advisory Board of SMTA (the Surface Mount Technology Association). He is a member of the ANSI-Accredited US TAG to ISO TC229 (Nanotechnologies). Dr. Rae holds a bachelor's degree from the University of Aberdeen and a Ph.D. and M.B.A. from the University of Newcastle upon Tyne. He is a member of the Royal Society of Chemistry and a Chartered Chemist.

Index

━━━ **F** ━━━